INVENT NOW

Salima Bhimani, PhD

INVENT NOW

The Next Leap in Moonshots

WILEY

Published by John Wiley & Sons, Inc., Hoboken, New Jersey.

For general information on our other products and services or for technical support, please contact our Customer Care Department within the United States at (800) 762-2974, outside the United States at (317) 572-3993 or fax (317) 572-4002.

Wiley also publishes its books in a variety of electronic formats. Some content that appears in print may not be available in electronic formats. For more information about Wiley products, visit our web site at www.wiley.com.

Library of Congress Cataloging-in-Publication Data has been applied for:

Hardback ISBN: 9781394394029
ePDF ISBN: 9781394394043
ePub ISBN: 9781394394036
oBook ISBN: 9781394394494

Cover Design: Wiley
Cover Image: © tortugin/stock.adobe.com
Author Photo: Courtesy of Celina Kenyon

Set in 12/16pt JansonTextLTStd by Lumina Datamatics
SKY10159927_062326

For my son.

For the world that is becoming.

And for those who taught me to create.

Contents

Preface

As we move into the next leap of moonshots,
this is meant to feel like ignition.
For you.
And for the world.
It may feel outside anything you—or the systems around you—have known.
That is not incidental.
Third-space moonshots emerge inside entanglements:
friction, competing incentives, inherited resource flows,
and possibilities that remain invisible when we look only where we've been trained to look.
This is not about standing at the edge, alone.
It is about moving into terrain where few will recognize what is forming—
and where those who are ready already sense what is at stake.
To build a moonshot is to move ahead of patterns when those patterns can no longer unleash what wants to emerge. Not to oppose them, but to out-create them.
Because the aim has never been to change one life.
The aim is to unlock opportunities that shape millions upon millions of realities.
This book is not an exercise in belief or optimism.

It is about accessing the collective intelligence already active in the world—and placing technology, capital, and imagination where unprecedented value and possibility can emerge.

If you are reading this,
it is not to remain where you are.
Or where the world currently is.
Disorient.
Grapple.
Be in what you were never told was available for you to lead.
This is orientation.

Introduction: Where Moonshots Begin

Five-year-old Jubeda hovered over a woodfire, flipping rotis to feed a family of 10. The air filled with that distinct smell of heat erupting wheat into orbital goodness. Rupa, her mother, had just returned from selling goods to seafarers making a pitstop at the shores of their village, Masindi Port, Uganda. Rupa had gone many times to the ships to ask about the needs of the mariners and travelers. She knew that understanding what they needed would help define the kinds of products she should make. And she kept asking questions so that each time her offerings were better and more useful.

But you see, she did not see her products as products. They were symbols of survival. Every pouch of spices sold or textile bought meant her family and by extension their one-lane village would have a day of food and money to buy resources for the community.

But this was not the first time Rupa, my maternal grandmother, had to reinvent herself. She and my grandfather had left Malia Hatina, a farming village in Gujarat, in the 1920s for what was ostensibly a better life.

Both sides of my family come from Gujarat, a western Indian state with a deep history of trade, industrial ingenuity, and entrepreneurial spirits. In the early 20th century, Gujarat adapted colonial-era machinery to build one of India's first modern textile economies. Today, it's home to a booming renewable energy sector, a thriving startup ecosystem, and the country's first international financial services zone. In recent years, the Gujarat startup ecosystem has drawn meaningful venture capital, with nearly ₹400 crore in funding flowing into startups.[1] But my family didn't benefit from that progress.

Both my paternal and maternal grandparents were small family farmers in Malia Hatina. They grew wheat and peanuts, whatever the land allowed, and lived at the mercy of unpredictable weather and volatile market conditions. Farming was labor-intensive, unforgiving work. So when the British colonial system offered a pathway to what was sold as a more "prosperous" path, my grandparents took it. They boarded boats under harsh conditions and migrated to Uganda in the late 1920s, missing, by mere geography and the timing of history, the grassroots innovations that would eventually emerge from Gujarat's soil—among them, nonhydraulic tilting bullock cart mechanisms that could have transformed agricultural labor had the lineage of invention and land remained unbroken.

My father was born in Malia Hatina, and his playground was the family farm. He completed high school, the farthest he would go with his studies. Growing up on a small farm was not for the faint of heart. Playing amid the tall vegetation, he would be confronted by the blazing sun, cobras staring him in the eyes, and parents who expected him to keep the farm going despite meager yields.

My Maternal Grandmother, the Original Innovator

Rupa never stepped foot in a school and neither did her youngest daughter, my mother. She raised nine children in a two-room mud house in Masindi Port, a village on the edge of Uganda where seafarers docked under the hot equatorial sun. There were no amenities. Meals were first cooked on open fires and then on a kerosene-fueled Primus stove. Everything was earned, built, or bartered by hand.

And yet, she was an innovator. She was a doula, a merchant, a spiritual advisor, a community matriarch. She delivered babies with her bare hands, mediated disputes, counseled families, and crafted goods. Her world didn't run on code or AI, but on intuition, adaptation, and systems of care built in real time, in the face of adversity and possibility—mingling together—with no blueprint.

As my mother always reminds me, they lived a simple, beautiful life. And it was not easy. My grandmother's gumption was born from necessity. She was curious, resourceful, and relentless. She absorbed what she needed to learn and turned that knowledge into opportunity, not just for her family but for her entire community.

A Legacy of Displacement and Reinvention

Eventually, my family realized that Masindi Port couldn't provide the future they hoped for. They moved to Kampala, the capital, where my parents opened a small café, Moez's Tea House. It was humble but promising. They still had no fridge, but at least they had running water and a city around them.

Then came the collapse.

In 1972, Idi Amin ordered the expulsion of all South Asians from Uganda. My parents, like tens of thousands of others, were forced to leave with just 50 shillings and a suitcase. Everything they had saved, including the money hidden under their carpets, was gone. The military ransacked homes. People were killed. And on the day they boarded a bus to the airport, my mother remembers the sounds of men, women, and children wailing—a deafening call of communal grief not just for being exiled but for being violently severed from the lives and selves they had shaped.[2]

My family scattered across the United Kingdom and Canada. My parents lived in the United Kingdom for three years before arriving in Canada where my story begins. They knew little English, but hard work and perseverance were their superpowers. I grew up with that energy. Amid the experiences of our family and our people. Not just experiences of trauma, but of resilience, imagination, and radical reinvention.

Not Just in Line

Have you ever stood in line at a restaurant and found yourself wondering about the people around you? Who are they? What do they do? How did they get here? What stories are silently unfolding just feet away from yours?

I was doing just that waiting for Indian food with Sergey Brin standing in front of me and Astro Teller right behind me. Astro Teller, whose grandfather co-invented the hydrogen bomb and who now leads Google X, the moonshot factory, and Sergey Brin, the son of a NASA scientist and a professor of mathematics and the co-founder of Google.

The Rails at Google X is no ordinary tech building. It's a place where ideas are birthed, killed, and sometimes launched into the stratosphere. Think Waymo (self-driving tech) and Wing (drone delivery). To find myself wedged between two of the most recognizable names in moonshot innovation felt electric, like I was standing in a current of unlimited potential. And then, a strange thing happened: the butter chicken felt more familiar than the presence of these two tech icons in line with me. I turned back, and Astro said, "Hello, Salima. What are you going to eat?" Sergey turned and smiled too. The conversation was all so mundane, even as the moment felt anything but. Suddenly it hit me:

> I am in line with these guys.
> *Damn, girl!*
> *And oh, God—I am in line with these guys.*

My stomach turned. Not from nerves, but because I realized the difference between a wide-eyed spectator and a moonshot maker isn't as big or as real as we've been told.

Calibration

Now, try this with me.
Close your eyes. Take a deep breath.
Imagine yourself standing in the heart of your city or town.
Ask yourself: how many lives can I change?
Picture one person.
Now five.
Ten. Twenty. Fifty.

A hundred. A thousand.
A hundred thousand. A million.
Five hundred million.
How far can your imagination stretch? How far can you keep the connection?
Can you see yourself shaping the lives of millions?
Can you feel the tingle of being in line with, not behind or beside, but *being with* the world's boldest inventors?

This is not just an exercise in imagination. It is a recalibration of your reach. You are no longer standing on the sidelines of the future; you are standing in the current where it is being authored. This book is the orientation for the work that follows.

What Are Moonshots, Really?

Fly me to the moon.

Let me play among the stars.

Let me see what spring is like.

On Jupiter and Mars.

Do me a favor: before you continue reading this chapter, go play Frank Sinatra's version of "Fly Me to the Moon." Close your eyes and listen. Come back to me when you're done.

This song drifted from radios in the late 1960s and later floated in space on the Apollo 10 and 11 missions.[3] It was a moment of cultural convergence, everyday hopes, and dreams merging with a cosmic, technological encounter. For me, it mirrors what moonshots are all about: being bold enough to imagine, to see, and to solve challenges from perspectives and places others might overlook or dismiss as out of reach.

I recall my first encounter with space. My neck was stretched back in a cushioned seat, as I sat among my fellow seven-year-old classmates on a school trip to the planetarium. An electric hum filled the dark dome. I closed my eyes, and as soon as the first collective gasp of awe rose around me, my eyelids snapped open. It was like seeing beauty for the first time. Stars, planets, the Milky Way, distant nebulae—they filled my vision. I floated in a cosmic oasis, my mind lifted into space and left to drift. It was just me and the universe orbiting each other.

As I sailed through solar systems, a swell of tears wet my baby cheeks. An eruption of emotions surged through me, feelings I couldn't yet name. All I knew was the overwhelming sense of being reborn.

When the experience ended, the electric hum returned. I felt myself descending from that cosmic suspension, dropping back to Earth and into my own thoughts. How did they do it? It felt like a miracle. How had they created this manufactured universe?

In the early 1980s, they would have used an opto-mechanical star projector—those mesmerizing machines with precision gears and rotating star fields that brought the cosmos to life inside a dome.[4]

The '80s weren't just about optical illusions of space. They were an incredibly active time for real space exploration. I remember my parents driving us to Florida for a family trip to Disneyland—just in time to coincide with a space launch. Though we couldn't see it from Kennedy Space Center itself, we watched from the outskirts. It was miraculous: the sky lit up, the rumble shook the ground, and the realization that humans could pierce the heavens settled into my bones.

Many years later, in my early 20s, I went to Jantar, an observatory created by Mantar Maharaja Jai Singh II in the 1700s. It was a massive stone observatory in India that measured time, predicted eclipses, and tracked stars with remarkable precision. Far ahead of his time, Jai Singh engineered these instruments to outperform many technologies from other parts of the world, making Jantar Mantar an early example of moonshot thinking in stone.[5]

My sense that we could innovate beyond what we know was rooted as a child and grew with me. But I never really had the language to tap into what a moonshot could mean or be. Astro Teller talks about moonshots in the following way:

> *"So a moonshot for us has three basic components. One, there has to be a huge problem with the world that you can name and you want to solve. Two, there has to be a radical proposed solution, something that's 10 times better than what's out there. Because if what you're proposing is either not a solution or is not radical, then why are we doing it? And then the third is there has to be some kind of core breakthrough technology that both gives us a reason to believe that that science fiction sounding product or service that you are imagining that could maybe solve that huge problem, could actually be built."*[6]

Alphabet's Moonshots

On January 7, 2025, Mike Johnson found himself trapped in a Waymo car circling endlessly in a parking lot. A video of the incident shows Johnson talking to a support person, asking, "Can't you just take control of the car?"[7] It's easy to look at that moment and think these are problems that should be immediately fixable. These tech companies should have all of this sorted out. From a user or customer experience perspective, that's an authentic and fair reaction.

But what many people outside the heart of self-driving technology don't realize is everything that led up to that moment Mr. Johnson experienced: 16 years of engineering to reach a point where a system can navigate thousands of complex scenarios and still sometimes fail. Yet in the moonshot space, this is part of the process. That's something I only truly understood once I got to Alphabet.

It was the spring of 2015, and I was asking myself one big question: *where can I contribute to the world now?* What industries are having the most dramatic impact on who we are as humans and what our experience of the world will become? Technology kept rising to the top.

Before 2015, I hadn't even thought much about Big Tech. But as I began reflecting on technology's economic and societal impacts—on both our present and our future—it became clear that this was where I needed to be. I wanted to help shape technology that could genuinely benefit more and more people and ideally reduce or eliminate potential harms along the way. The Internet. Navigation maps. Cell phones. Chat bots.

And I realized something else: I'd never worked in an industry where the combined market capitalization of its largest players often exceeds the total GDP of entire continents like Africa or South America highlighting the scale of economic value investors expect these companies to create.[8] Of course, my own "valuation" of Big Tech wasn't just based on investor perspectives, but it put into sharp focus the opportunity to influence technologies that are increasingly ubiquitous in our daily lives.

So I did what many of us do: I looked on LinkedIn. The moment I clicked, a job at Google popped up. It looked interesting, so I applied not thinking too much of it. Twenty-four hours later, I got a ping asking if I'd be interested in an interview. Still not making too much of it, I thought, *Sure, why not?* You see, I didn't grow up dreaming of working for Google. What I really wanted was to be somewhere I could have a high and real impact in shaping the world.

After my first interview with the recruiter, she said, "I think there's another role for you given your decades of experience. It's a brand-new leadership position with a scope that's yet to be fully defined, but it would involve working across the Other Bets companies to help them achieve their missions."

I hid my inner excitement, but this was the leap I was looking for. I want to shape missions, leaders, technology, and businesses from the ground up. I want to 10X their 10X ambition.

I'd be part of the Alphabet ecosystem, with the resources of Big Tech, but working with their portfolio of startup-like companies. Seven interviews later, I was hired and became Alphabet's first chief strategist for systemic change to work with its Other Bets portfolio.

I came into a thriving, Other Bets ecosystem. On August 10, 2015, Larry Page and Sergey Brin announced that they were creating Alphabet, separating Google from their moonshot projects. In their letter about this separation, they wrote the following:

> The company evolved into Alphabet in 2015. As we said in the Alphabet founding letter in 2015. *"Alphabet is about businesses prospering through strong leaders and independence."*

> Since we wrote that, hundreds of Phoenix residents are now being driven around in Waymo cars—many without drivers! Wing became the first drone company to make commercial deliveries to consumers in the U.S. And Verily and Calico are doing important work, through a number of great partnerships with other healthcare companies. Some of our "Other Bets" have their own boards with independent members, and outside investors.
>
> Those are just a few examples of technology companies that we have formed within Alphabet, in addition to investment subsidiaries GV and Capital G, which have supported hundreds more. Together with all of Google's services, this forms a colorful tapestry of bets in technology across a range of industries—all with the goal of helping people and tackling major challenges.[9]

These were projects that Larry and Sergey had personal interest in. Some of these projects were being birthed out of X, the moonshot factory, which was launched in 2010, as the secret R&D lab. Some, like DeepMind, were acquired. And others, like their VCs, also grew out of Google, as part of the Other Bets portfolio.

Joining this ecosystem was overwhelming and completely fascinating. I was excited every day. I had found myself in a place of possibility.

The Third-Space Moonshots

At this point, you might be wondering, what is this book really about?

Let me be clear. This is not a love letter to Silicon Valley moonshots, and it's not a takedown of them either. Both positions are too small. Moonshots, by definition, are meant to expand what we believe is possible. Writing from either pole would miss the point. I'm not here to rehearse what we already know. I'm here to surface what we haven't yet learned how to see.

When I arrived at Alphabet, I was inside the room where moonshots were imagined and resourced. It allowed me to see moonshots not simply as high-risk ventures, but as a distinct architecture of possibility. From inside that architecture, I came to a realization: traditional moonshots, while ambitious, aren't ambitious enough.

They don't lack brilliance. They have produced extraordinary breakthroughs—AlphaFold, robotics, AI, advances that matter deeply. But could they be more if they came from an expanded set of tools and vantage points?

This book is about the emergence of a third space—a new generation of global moonshots that represent the next leap in how we invent. These moonshots don't live in technological bravado alone, and they don't emerge from grassroots ingenuity in isolation. They arise at the convergence of both.

We are entering a period of history where disparate realities—the village and the lab, the ancestor and the algorithm can finally converge. The infrastructure of the world is becoming porous. The "Rails" of high technology are reaching the hands of the merchant in Kampala and the farmer in Gujarat. A moonshot without lived intelligence is blind, and lived ingenuity without scale is a tragedy of untapped potential.

The future belongs to moonshots that do something more demanding: they converge the ingenuity of everyday life with the precision of advanced technology. These are not abstractions lifted away from context, but interventions grounded in lived systems—cities, villages, markets, institutions, relationships, and cultures—and then amplified through technology capable of reshaping entire ecosystems. This is no longer just an option; it is our most urgent mandate.

Why Now?

I once had a conversation with an engineer about drone deliveries. He told me, with the earnestness common in the lab, that the goal was simple: "We are going to give people more access to things so they can focus on *other* things."

It is a clean, logical promise. There is something intriguing about the idea of giving people more space to imagine and create. I find myself drawn to it. While simultaneously it is the gospel of efficiency. So as I listened to him, I thought about the "other things" people are actually focusing on today. For most of the world, those "other things" aren't creative hobbies; they are the deep, vibrating vulnerabilities of social, economic, and technological instability echoing across the world.

Most people I speak with describe a similar sensation: a growing sense of ambiguity about the future and tension in the present. The average person doesn't experience the "promise" of tech as a gift of time; they experience it as a tidal wave of change that threatens what they know while presenting opportunities they cannot yet grasp.

We have an intuition that the future is arriving unevenly, all at once. Some people cling to what was; others rush toward what might be. Many find themselves suspended in between. We are encountering a crash of forces unlike anything in human history—a civilizational hybrid. This shift sits in the pit of the stomach. It's the question of whether our skills will matter in five years or how we will feed our families if the ground beneath the economy continues to shift.

These are valid, heavy questions. But there is a more precise question we must ask: what will we choose to create in the midst of a change we cannot fully predict?

This is more than a technological shift; it is an evolutionary one. We have arrived at a point where we are no longer just builders holding tools; we are emerging as a synthesis of biological intuition and planetary-scale processing. We are becoming a civilizational hybrid.

This isn't about sci-fi abstractions. It's about the fact that our nervous systems and our networks are now integrating circuits. The way we think, connect, and solve problems is now happening through a porous boundary between the village and the lab. We are the first generation whose ancient survival instincts—the "Roti"—are being processed through the "AI" of our collective intelligence.

Earlier in this chapter, I asked whether you can feel yourself shaping the lives of millions of people. That question lives at the center of the moment we're in, as individuals, as societies, and as a collective intelligence.

In a hyper-interconnected world, every architecture we build, every switch we flip vibrates through systems of every scale. That "low-grade hum" beneath daily life isn't the sound of a crash; it is the vibration of a massive, global integration. It is happening through the neural pathways of artificial intelligence and the silent leaps of quantum computing. We are the first generation of builders who don't have to choose between the Roti and the AI. We are learning to hold multiple truths at once.

The question is no longer *if* technology will shape our future. It is whether we have the audacity to make it a partner in a new kind of inventing: one that is as grounded as it is galactic.

Why the Leap Is Necessary

For the first time in history, we have the potential to meet the complexity of our global systems and dynamics with technology in a way that simply was not possible before. This isn't a temporary trend; it's the new baseline.

The traditional moonshot space has always been defined by a refreshing, almost defiant ambition to build solutions never imagined before. But for too long, that ambition has been narrow. The spark required to meet global-level complexity belongs to a much larger set of people.

The current model is insufficient because our challenges have shifted from the complicated to the interdependently complex.

The next frontier of 10X'd 10X growth sits in the complex. Complex systems—the ones that govern how the world eats, moves, and builds wealth—cannot be solved in a lab alone because a complex problem is alive. It is a shifting web of culture, geography, and human behavior. Most builders avoid these systems because they are messy. They see a rusted switch and walk away.

But that mess is actually the signal of a massive, latent demand. A mess in an ecosystem is possibly where intelligence is untapped. When you see a village where farmers can't get their goods to market, or a city where waste isn't being converted to energy, you aren't looking at a problem to be helped. You are looking at a market that could come alive.

The leap is necessary because we can finally bridge the gap. We have the chance to birth a third space of possibility by taking the ingenuity of billions of people, the forces of systems, and the radical engine of technology.

The leap is to start building in places of latent opportunity.

Moonshots Shape Billions of Lives—but Billions Rarely Shape the Moonshots

Think about one of the most consequential technological integrations into modern life: Google Search. More than five billion people use it worldwide. It has reshaped how we access information, how we think, and how the world becomes available to us moment by moment. It is, without question, a moonshot that has altered the lives of billions.

Yet, almost all of those billions experienced that transformation as receivers. Few were involved in shaping it.

This raises a deeper question: where will the next world-shaping opportunity emerge and who will build it?

Will it remain the domain of a handful of institutions and companies? Will the rest of us simply receive the consequences of someone else's vision?

Or will more of us—you, your cousin, your community, your city, your country—step into shaping the forces that redefine how the world works?

The Invitation to Invent

The moonshot practice I invite you into is not a series of stand-alone companies, startup ideas, or isolated bets. When you begin to see them as ecosystem-level interventions, their true scope comes into view.

What if the most powerful moonshots haven't been built yet—not because the technology doesn't exist but because we've been confined by inherited assumptions about who gets to invent? What if the people capable of building them haven't yet realized that they can, and must?

So let me be precise: the third space is not a compromise. It is not Silicon Valley "light," nor is it innovation from the margins seeking validation. It is a new architecture altogether. Third-space moonshots don't sit between worlds; they emerge from an open field where multiple logics exist simultaneously for our orchestration.

It is where ambition expands because accountability expands—to histories, to futures, and to entire ecosystems of people and possibility. We are not here to build "point solutions." We are here to unlock entire ecosystems of opportunity.

This is not a theory of innovation. It is a practice of inventing.

Inventing now.

Because the next generation of moonshots isn't arriving in some distant "someday." It is already forming. We need it more than ever. And it needs you.

The Orientation

We don't follow a map. We learn how to orient ourselves to moonshot making. This is not a how-to manual; it's an orientation. A challenge. An invitation to rethink what moonshots could be—and what it means for you to build one.

In the chapters ahead, I'll share stories from my time at Alphabet and far beyond it. I'll offer ways of seeing, mental models, and practices designed to help you step fully into the role of moonshot maker.

This will be a dance.

It won't always be smooth. We'll misstep. We'll get things wrong. That's not a failure of the work—it is the work itself. Moonshots don't require perfection. They demand presence, imagination, and the willingness to stay with complexity long enough for something new to emerge. We are not here to measure success and failure by inherited metrics. We are here to stretch what we know into what the world is asking for next.

You don't need permission to build what only you can imagine. And you don't need permission to imagine what you haven't yet. You already carry the intelligence of necessity and the spark of invention. This book will orient you, not by giving you answers but by tuning you to the rhythms, approaches, and practices that make moonshots possible.

So take a breath. Drink in some audacity. Step into the space between what's known and what's waiting to be discovered.

Not because you've arrived. But because you're ready to launch.

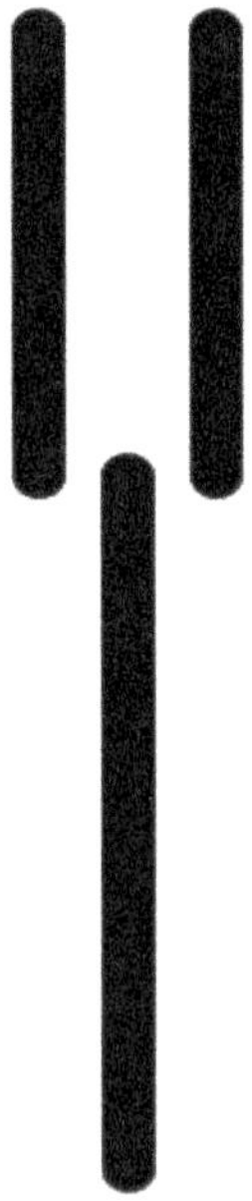

CHAPTER

1

Moonshots Are Living Experiments

I fell in love with theater in seventh grade, at 12 years old. My middle school was staging *Oliver Twist*, a play about a young orphan finding his way through Victorian England. I auditioned and got the part of Nancy. She didn't have many lines, but I didn't realize then that I was stepping into one of the play's pivotal moments—Nancy confronting Fagin, the villain.

I practiced endlessly: in rehearsals, in my bedroom, in front of the mirror, shaping every word, trying to feel what Nancy wanted to say. On opening night, I peeked through the curtain and froze. The audience stretched out before me—row after row, what felt like hundreds of eyes. My stomach dropped. My breath caught. I bolted back to my mark, heart pounding in my ears.

When my cue came, I whispered to myself: *I am Nancy*. I walked on stage, locked eyes with Fagin, and began my lines. At first, I clung to the script, each word carefully rehearsed. But then something stirred. A pulse of energy surged from my belly, rose to my throat, and when he delivered his final line—

I erupted.

"NO!"

It ripped through the theater, raw and unplanned. The audience gasped. Silence followed, heavy and electric, before it broke into an uproar—applause, cheers, people rising to their feet. The play stopped cold for two full minutes.

That night, I discovered something I would carry with me for the rest of my life: the power of breaking the script. The power of improvisation, of trusting the moment, of letting ingenuity erupt into something bigger than yourself—and ripple outward to impact not one person but hundreds.

What if every moonshot began by going bigger than big?

What Is a Moonshot?

It had all the makings of a technological marvel. Engineered for durability in harsh environments. Designed with modular components for easy maintenance. Capable of extracting a critical natural resource from deep below Earth's surface. Scaled across thousands of sites. Trusted by millions.

You could mistake it for a piece of off-grid survival tech from a sci-fi film. But it wasn't hidden in a secret lab or launched with a billion-dollar budget. It stood quietly at the center of rural villages across Zimbabwe: the Bush Pump.

Made of steel and mounted on concrete slabs, the pump used a simple lever to draw water from aquifers. With each pull, clean water rose into buckets, basins, and clay pots. By the 1970s, Zimbabwean engineers and public health officials had reimagined the original colonial-era design into what became known as the "Type B" Bush Pump. Tough. Repairable. Built from local materials.[1] And—most importantly—maintainable by the very people who used it.

Its simplicity was its brilliance. No specialized parts. No imported expertise. A farmer could fix it. A teacher could manage it. Local workshops could fabricate components, and community groups oversaw installation and upkeep. Over time, tens of thousands of Bush Pumps were installed across rural Zimbabwe, becoming a nationally standardized approach to water access.[2] The pump's durability, adaptability, and local maintainability allowed it to endure for decades—not as a static invention but as a living system shaped by the people who depended on it.

The Bush Pump wasn't a prototype or a pitch. It was national infrastructure. A piece of engineering that quietly achieved what so many futuristic inventions promise but rarely deliver: improved lives at scale.

So is the Bush Pump a moonshot?

For many of you, the idea of a "moonshot" might feel distant, even unfamiliar. But it doesn't need to be. Think for a moment: what technologies in your own lifetime have shifted what you believed was possible? Which ones changed not only how things worked but how life itself felt?

The Bush Pump reminds us that transformative innovation does not always come from billion-dollar labs or venture capital pipelines. Sometimes it emerges from local ingenuity, built with whatever resources a community can muster. That same spirit of turning ordinary spaces into launchpads for the extraordinary is what I encountered in Silicon Valley.

Laying Down the Tracks for Moonshots

In 1963, the Peninsula Curling Rink was a hub for winter sports. By the early 1970s it had been transformed into Mayfield Mall—one of Northern California's first air-conditioned shopping centers.[3] Just down the road, Stanford, Fairchild, and Intel were already stirring with a new kind of ambition, but this site felt ordinary. Threshold space. Waiting.

Decades later, that same structure would become home to some of the world's most audacious technological experiments: self-driving cars, drone delivery, AI-powered agriculture. Yet the building never fully shed its past. The ceilings remained high, the walls unfinished. You could still feel the bones of the mall beneath the concrete and steel. It wasn't built to be a lab—it became one.

That's what I felt walking into the Rails, home to Alphabet's "moonshot factory." The metal scaffolding, the robots, the laptops glowing with product roadmaps—all of it radiated a particular kind of ambition. Engineering benches doubled as aesthetic statements. The space pulsed with minds bent on achieving the unimaginable.

If learning about the Bush Pump taught me that moonshots can emerge from the ground up, the Rails taught me that they could also erupt from the top down when resources, talent, and imagination are condensed into a single, restless place. Both were systems in motion, waiting for someone to keep the beat.

A year after I got to Alphabet, I was sitting with John Krafcik, then CEO of Waymo. Behind him, mechanical levers moved in quiet rhythm while nearby engineers argued over the math behind a stubborn problem. Just as my thoughts wandered, his voice cut through with clarity: "Salima, at Waymo, we're not building self-driving cars. We're building technology that redefines mobility, across systems, across contexts. A self-driving car is a worthy goal. But our target is bigger: technology that scales."

That word *scale* stayed with me. Scale is not just about size; it's about rhythm, the steady repetition of a focused yet elastic idea that oscillates until it reshapes technology that redirects the world. That's what Silicon Valley has cultivated since its earliest days—not just tools but transformation. From HP's garage in the 1930s to Google's in the 1990s, the startup mythos has endured. Scrappy beginnings, world-changing scale.

In my farewell note before leaving Google, I wrote: "How many people get the chance to bring their most audacious ideas to life with a company that literally changes the lives of billions of people? I can now say that I am one of them."

That truth is what made leaving so charged. Alphabet is not just a company—it is an entity. A living experiment in how collective intelligence can shape humanity's future. To step away was to feel as though I was stepping off the tracks of history as they were being laid. I still wonder if Larry and Sergey knew what they were unleashing in the early days—that the boldness of their idea would ripple outward into something with unprecedented implications. Did they then know that search would get approximately 5 trillion queries a year?[4]

The Tabla Rhythm: The Discipline of the Loop

Inside Silicon Valley's most ambitious labs, it can feel like moonshots erupt in a single flash of genius—a breakthrough algorithm, a successful test flight, a perfectly timed pitch. But that's not how it really works.

There's a rhythm to moonshot building. A discipline that unfolds over time.

Think of Indian classical music—specifically tabla. Listen to this: (Pt Swapan Chaudhury Teental):

At first, its 16-beat cycle ("teen taal") sounds like simple repetition: a steady wave of time. But within it lives something more powerful: structure, improvisation, and return. Every beat has a role. Even the silences have weight.

Moonshots follow the same pattern. They aren't a single explosive leap into the future. They unfold in cycles—moments of tension, experimentation, correction, and resolution. They demand both vision and repetition. Both ambition and timing.

And like a master tabla player, moonshot builders must learn how to stretch and bend without ever losing the rhythm that carries them forward. Moonshots, like tabla, are not about a single downbeat of invention. They are about the rhythm that makes transformation possible.

This rhythm reveals why moonshots can't be understood as isolated and single point inventions. They are less like solitary notes and more like movements in a composition, each shaping the other. Which is why Alphabet's Other

Bets should not be seen as scattered ventures but as an ecosystem—a set of interwoven beats directing the larger trajectory of human possibility.

More Than High-Risk Ventures

I like to think of Alphabet's moonshot companies not merely as a portfolio of high-risk ventures but as a living ecosystem of emerging technologies. From a traditional business perspective, it makes sense to view them as a pipeline for the next billion-dollar company—the next generation of scalable solutions to urgent and unforeseen problems. But reframing them as an ecosystem allows us to see something deeper. The "Other Bets" are not simply investments; they are inputs that redirect human trajectory.

In this view, these companies become more than groundbreaking technologies or mission-driven businesses. They act as architects of human experience and human evolution. Each was birthed or brought in under the infrastructure of Alphabet, rooted in Google's success. They were funded year after year, granted access to Google's vast data and code base, and supported by organizational systems for hiring, business development, and operations. Like saplings under a giant canopy, they benefitted from soil already enriched by Google's growth.

Yet that nourishment came with tension. As much as the Other Bets were Alphabet companies, they were also expected to become their own entities—each with its own culture, identity, and trajectory. Most of their CEOs did not come from Google and worked hard to establish independence. At the same time, Googlers who transferred in carried expectations of Googleyness, pressing for familiar ways of working. The result was a constant push and pull between dependence and autonomy, between inherited DNA and the desire to mutate into something new.

But ecosystems are never tidy. They carry both nourishment and friction. Inside Alphabet, every Bet was shaped not just by funding and infrastructure but by the messy work of people trying to define who they were and who they were not. Some leaders embraced the privilege of Google's canopy while straining against its shadow. Others insisted that independence meant repeating the very patterns they hoped to escape. These tensions are where the ecosystem comes alive in the stories of leaders, engineers, and cultures wrestling with what it meant to grow a moonshot. To me this was the very necessary act of not just shaping a moonshot but living its ethos.

Dan Doctoroff, the CEO of Sidewalk Labs, and I were digging into this in one of our regular one-on-ones. He spoke candidly about this paradox. For him, the struggle to be their own company while still reaping the benefits of Google and Alphabet was both confusing and a privilege. It made the work of building distinct Bet cultures and systems all the more challenging. And yet, as he noted, these companies would never survive the outside realities of funding and growth expectations. The runway they had inside Alphabet was extraordinary—more than unique. The executive team at Sidewalk Labs worked hard to craft their own path, while working through the tension deliberately to not be Google.

This view was amplified by leading engineers at Waymo but from the opposite perspective. One day in a meeting, I had shared reflections on the pitfalls Google itself had faced in trying to sustain a culture of innovation, how it wrestled with not being a startup, while still wanting a startup ethos, and how what the company often said did not always bear out in employee experiences or expectations. I suggested they had a chance to learn from those lessons. But their response surprised me. As soon as I shared this, the tension in the room rose. "Salima," they said unanimously, "this is not true. We need to emulate Google." The irony was striking: even as I urged them to move beyond Google's contradictions, they insisted on reproducing them. It revealed how difficult it was to apply a learn-and-grow ethos inside the very ecosystem where Google's DNA remained dominant.

Seen as an ecosystem, the Other Bets reveal how moonshots are not just launched but grown, nurtured, contested, and sometimes stunted—within a larger environment. Like organisms in any ecosystem, they must find their way from fragile seedlings to maturity or risk being absorbed, reshaped, or cut off.

Take Isomorphic, for example, a company that emerged out of DeepMind, Alphabet's AI powerhouse. Its breakthrough with AlphaFold transformed the "what, how, and when" of drug discovery. Suddenly, the long arc of medical research bent forward, accelerating timelines, collapsing uncertainty, and opening pathways for treatments that once seemed out of reach.

AlphaFold's release of 200 million protein structures was not just another dataset.[5] It was a shift in possibility. Malaria, genetic diseases, antibiotic resistance—challenges that once required decades of painstaking research could now be addressed with a precision and speed previously unimaginable.

More than a tool, AlphaFold became an ecosystem-level unlock. It catalyzed an entire wave of life-science research and discovery, enabling breakthroughs well beyond its own lab. Moonshots like these do not remain contained; they ripple outward, reshaping what is possible for human health.

From the outside, moonshots like AlphaFold look like breakthroughs. From the inside, they feel like a rhythm—a culture, a cadence, a lived experience. This rhythm wasn't limited to AlphaFold. It pulsed across Google itself.

When I arrived at Google in 2017, people told me I was joining at the tail end of "old Google." I had no baseline, but there was a distinct vibration of what had been. Google was the size of a small city, so not everyone experienced it the same way, yet the undercurrent was clear: here, anything felt possible. Ingenuity, hard work, and grit mattered. You could chase ideas, join new projects, or learn from other teams—the pathways were open. Technology, radical solutions, and ambitious research weren't side projects; they were the heartbeat.

Of course, the stereotypes of big tech were also true. Free meals, endless snacks, generous health benefits, laundry services. I never saw these as catnip, but as part of a deliberately curated culture. Entering it meant you were somewhere distinct, a place of abundance that few other environments could match. Yet as the company scaled, it also had to operate like a corporation: people needed to feel seen and heard, while remembering this was still a business.

The ethos said hierarchy didn't matter and every voice had value. In practice, it was more complicated. The Googleverse was shaped by rank, seniority, and access, even as it amplified horizontal collaboration. The vertical was played down, but it was there—sometimes slowing progress when 10 people had to weigh in on a single document. And yet Google continued to grow and develop with momentum.

In my first weeks, I didn't know the Other Bets world was different. Over time, I came to be grateful that my daily work was a field of play. If the solution meant speaking directly with CEOs of the Other Bet companies, I could. If it meant spending hours with senior engineers, I could. If it meant bringing ideas no one had considered, I didn't need months of approvals to test them. The Other Bets world was one of audacity, play, and experimentation—a rhythm of ingenuity unburdened by corporate bureaucracy. It gave me the space to lead, to grow, and to invent.

And perhaps more than the freedom I felt to innovate with the Other Bet companies and their leaders, what moved me most was the people themselves. They cared deeply about their missions and the hard problems their technologies aimed to solve. Everyone I worked with across the Other Bets had joined because they wanted to leave an impact on the world. In the end, the most powerful resource in the ecosystem wasn't capital or code. It was human energy—the conviction of mission-driven individuals who believed their work could bend the trajectory of the future.

The Unique Setup

What made the Other Bet companies distinctive is not only their ambition but the infrastructure that gives them form. Many big tech companies have moonshot-like efforts—such as Amazon with Prime Air drones and Microsoft with its research divisions and partnership with OpenAI. But these live as initiatives within a larger corporate body. Alphabet has done something no one else dared:

it has carved out a separate architecture for moonshots themselves.

The Other Bets are not side projects; they are companies, each with its own CEO, PandL, and brand. They are given space to breathe, to scale, to die, and to be reborn. Within this ecosystem, ideas can be nurtured, tested, or killed outright, while new ones are continually seeded. The moonshot space has its own mythos and motivations, grounded in the drive to 10X, to leap beyond incremental improvements into breakthroughs with the power to reshape industries and lives.

Alphabet, in this sense, built something larger than an incubator. It's built an institutionalized portfolio, a deliberate long-term investment in creating entirely new companies for the world. That choice was and continues to be radical. Alphabet reframed moonshots from occasional bets to an organized system of invention.

It's easy to focus just on what the companies do given the scope of these companies. But we really get a better sense of them when we uncover the areas of life they are evolving.

While Alphabet's bets are often discussed individually or by sector, their deeper coherence emerges when viewed through the human systems they collectively seek to reshape. The following table reflects my synthesis of how Alphabet's major bets function as ecosystem-level interventions—each targeting a distinct dimension of human experience yet operating together as a coordinated whole. Some of the examples included represent bets that were later wound down, underscoring how experimentation, learning, and closure are integral to how this ecosystem operates. And today the Bet ecosystem is evolving even further.

Alphabet Bet	Domain	Human System Being Re-architected
Google	Digital infrastructure	Knowledge, communication, work, commerce, and everyday digital life
X (Moonshot Factory)	Invention	How radical ideas are generated, tested to become the next Bets
Waymo/Wing	Mobility and logistics	Movement of people and goods; access, delivery, and spatial efficiency
DeepMind/Isomorphic	Intelligence and science	AI development and adoption, scientific discovery, knowledge creation, and problem-solving at planetary scale
Verily/Calico	Health and longevity	Precision health, disease intervention, and the science of longer lives
Intrinsic (moved to Google)	Robotics	How physical work is automated, augmented, and integrated into manufacturing and industry
Google Fiber	Connectivity	Broadband infrastructure and access to high-speed digital networks
Loon *(sunset)*	Global connectivity	Internet access in remote and underserved regions via nontraditional infrastructure
Sidewalk Labs *(sunset)*	Urban systems	How cities integrate technology, governance, infrastructure, and public life
GV/CapitalG	Capital	How innovation is financed, scaled, and sustained over time

Alphabet's Other Bets ecosystem is not just a collection of experimental ventures. It is also a flow of capital, talent, and imagination designed to ripple outward into society. GV alone manages close to $10 billion across startups,[6] while CapitalG, Alphabet's growth-stage arm, oversees another $7 billion, channeling resources into scaling bold ideas.[7] Waymo, once only a sketch of autonomous mobility, now carries out hundreds of thousands of rides each week,[8] with ripple effects from safety gains to an estimated $40 million a year in added local tourism activity in San Francisco.[9] DeepMind shifted the trajectory of AI itself by embedding intelligence directly into the systems we rely on: its models reduced energy use in Google's data centers by up to 40%[10] and increased the value of wind power by 20% by predicting output 36 hours in advance.[11] Intrinsic, Alphabet's robotics venture,[12] is entering manufacturing just as the global robotics economy is forecast to grow from approximately $74 billion in 2025 to $373 billion by 2035.[13] Intrinsic has now folded into Google contributing greater leverage in Google's AI and robotics pursuits, enabling factories to cut costs and reimagine production. Seen together, these companies are instruments reshaping how we move, heal, make, and imagine. An ecosystem that acts as architecture for human experience.

What made this ecosystem even more interesting is the life cycle of opportunities that Alphabet created to bring new Bets into the world.

Take X, for example. It had a full process for seeding and researching ideas that could yield real technological solutions, impact millions of people, and potentially become billion-dollar businesses. Some ideas die early because they don't meet the test. Others evolve into incubated projects, like Everyday Robots or Taara. And still others grow beyond projects into fully graduated companies like Wing. Just within X, you see an engine of innovation in motion.

Alphabet also grew by acquisition, most notably DeepMind, which reshaped Alphabet's place in the AI landscape. Across the portfolio, Bets live at different stages of maturity: from raw experiments to companies with established market penetration and business goals. The ecosystem is alive, shifting constantly.

Critics might point out that far fewer Bets have created lasting economic or business value than expected. But this misses part of the design: moonshot lifecycles are meant to be hard. They are given extraordinary runways precisely because developing breakthrough technology takes years. In this model, technology comes first; business models and go-to-market strategies often follow. I often saw this as a *slowing down* of potential moonshots to become companies in the world with real impact.

That dynamic has shifted. Under today's economic pressures, the Other Bets are pushed toward market viability much earlier. The return on investment must now be measured in revenue, not just technological breakthrough. Bringing in external funding is part of the matrix now. Some of the early spark of unfettered experimentation has dimmed. But in another sense, this pressure forces sharper evaluation: *Where does this technology truly fit in the world? Will it make money?*

The challenge, of course, is that many of these evaluations rely on traditional market metrics that don't always fit the radical nature of moonshots. As one CEO put it to me in a candid exchange echoed by engineers across various Bets, "We are not trying to reach all markets. We must, and should, always start with the markets where the money is. It's a matter of survival."

From my vantage point, the Other Bets were designed to thrive because they are rooted in a distinct set of conditions:

- **Massive funding** from Alphabet to get off the ground
- **Time and space** for ideas to incubate and mature
- **Shared resources** like AI capabilities, code, systems, and people
- **Talent flows** from within Google and top recruits from outside
- **Protection** under the Alphabet umbrella
- **Guidance** from Larry, Sergey, and other key executives
- **Friction and tension** of being independent companies while still in the Googleverse
- **Start-up nimbleness** without the same existential pressures of true startups
- **Mission-driven leaders and employees**, deeply committed to the work

These are the roots of the Other Bets ecosystem—roots that allow for speculative ideas with the potential to become the next Google and to generate real impact in the world. In some ways, the roots opened new technological pathways to address urgent problems. In other ways, those same roots limited the companies from moving beyond a single 10X to 10X'ing their 10X opportunity.

This is where the moonshot process of red teaming comes in: asking what could truly take these Bets to the next level. For me, this was the opportunity—the chance to push their 10X ambition into a new flight path.

Looking back, the Other Bets were designed to thrive because they were rooted in a distinct set of conditions: massive funding, protected time, and a canopy of shared resources. But as an insider standing at the intersection of these moonshots and the world they were meant to serve, I began to see how the architecture itself shaped the output.

It wasn't a lack of vision; it was the internal logic of a high-velocity ecosystem. When you are calibrated for 10X, the "standard" becomes a powerful gravity well. I began to track the patterns that emerged from this particular microclimate:

- **The proximity filter:** Our starting points were naturally rooted in North American markets. It wasn't an intentional exclusion; it was a design reflection. When your data, your users, and your feedback loops are tuned to a high-bandwidth reality, you naturally build for the world you can touch.
- **The logic of the smooth(er) track:** We optimized for "clean" contexts where our technology could run at its highest velocity. This inadvertently bypassed the "unpaved" realities where the need—and the potential for resonance—might be even greater.
- **The tech-first pulse:** In the Rails, the breakthrough was the sacred event. We treated technological solutions as the most valuable starting point, often expecting the context to adapt to the tool rather than the tool evolving to meet the human ecosystem.
- **The canopy abundance:** Our extraordinary runway was a gift, but it also functioned like a silencer. It was difficult to hear the "ripples" of real-world friction—the local costs, the cultural nuances, the ethical weight—when we were protected by the most sophisticated corporate architecture ever built.

These patterns weren't "failures." They were the practiced settings of a particular moonshot approach. To move beyond a 10X to 10X'ing the 10X opportunity, we don't need to discard this approach. For those who can and want to moonshot we need to widen the aperture and ask, what else is possible?

Multiplying Force

The Bush Pump comes back to mind here. It did not emerge from a billion-dollar lab, yet it rippled through an entire nation. But we have to ask: could the Bush Pump have been even more impactful with other technologies at hand, such as remote sensing for aquifers or decentralized energy for the lift?

Alphabet's Other Bets, by contrast, are infused with resources, talent, and ambition, yet often stay bound to a narrow frame. They have the "galactic" reach but often miss grounded resonance. Could they be more if they tapped into the intelligence of the street?

This is more than a tension worth holding; it is a site of eruption. It is the space where ingenuity that rises from the ground up meets the moonshots born from the top down. It's about the centrifugal audacity—the human and technological possibility of birthing something else.

We aren't looking for a compromise. We are looking for the multiplying force that happens when ingenuity and moonshots occupy the same trajectory. The question is not whether one is better than the other. The question is whether we can imagine a space where they don't just coexist but co-evolves.

If you were standing on the stage of possibility, how would you shape this **third space**? Would you settle for 10X, or dare to take it further?

CHAPTER

2

The Moonshot Mythos

My father would come home after a 12-hour workday. Yes, he was, in some ways, the classic Indian immigrant stereotype: a man running a small convenience store in an apartment building in uptown Toronto. But it was there, in that modest business, that I learned my first lessons in entrepreneurship at ten: how to choose products for a specific customer base, buy inventory and track costs, and build simple spreadsheets to calculate overhead and profit.

Each night, he'd drop thick wads of cash onto our kitchen table and ask me to help. I'd separate the bills by denomination, stack them into neat piles, count them one by one, and call out the totals. Holding that much money stirred a quiet wonder in me—a sense that resources, even modest ones, could build things, could change things.

As I sat beside him night after night, my 10-year-old mind wandered. What could I create with this? A cure for disease? A rocket ship to take kids to space? A machine to turn pollution into clean air?

In the end, the cash went to paying bills—to keeping our family safe, housed, and fed. But the feeling it sparked never left: a restless curiosity about what's possible when you have something to build with.

I wonder if those nights at the kitchen table quietly seeded my belief that I could create anything. Back then, I didn't think about what it takes to make an idea real—or the barriers that stand in the way. I just had that belly fire, the steady sense that ingenuity and innovation were my birthright.

Years later, when my own ideas began to whirl into the world, I understood something I couldn't have known as a child: innovation can feel impossibly out of reach for many people—and for others, easily accessible.

I was never fascinated by the figures of innovation. I didn't spend much time thinking about Bill Gates, Steve Jobs, or Jensen Huang. I cared more about the impact their sparks had on the world. Maybe I sensed that if I stared too long at the people behind the innovations, I'd feel further from the work itself, as if their stories were so distant from mine that, if I wasn't careful, I'd start to believe it wasn't for me.

Not everyone needs or wants to be an innovator. And yet what you're exposed to, what you're part of, and what you stumble upon or survive quietly shape what you think is possible. For many, the moonshot story isn't about what can be done, but about what they've been shown is for them.

The Friction of the Mythos

When I think about the moonshot mythos, I find myself holding mixed feelings. The story is not as simple as the stereotype of big tech and lone geniuses. There is more happening beneath the surface than the narrative usually allows.

That is why I find the moonshot mythos both boring and compelling. A mythos is required when you want to do something big. It isn't just a story we tell; it's a system we live inside. It names our heroes, edits our failures, and turns ambition into action. Every industry has one; the moonshot world just builds its mythos at planetary scale.

What made it compelling at Alphabet is that as you enter the Other Bet world, it feels like a place you belong to by default. My manager would introduce me to people with a proud gravitas. My experience and my background were treated as a necessary addition to a place where innovators were not just welcomed but expected. It was easy to feel part of something bigger—that you were necessary to the Mythos, too.

But life in that space was more of a dynamic friction than a stable truth.

At Alphabet, the mythos orbited a familiar constellation of forces: the founder-visionary, the technical prodigy, and the operator who could move reality just enough to make the initially unthinkable plausible. But the myth wasn't only about the people; it was about the idea of the moonshot itself—that scale, audacity, and disruption were the truest markers of progress. Over time, "moonshot" stopped being a metaphor for possibility and became a category of belonging, a way to decide which ideas deserved to exist. The stories, the stage time, and the capital were built into the center of these forces. The myth mapped its stars in this constellation.

But the mythos also, at times, was boring—because it's the story we've been told for decades: the few geniuses, the improbable vision, the destined path, the game-changing tech. This predictability isn't accidental; it's built on where you sit and which doors open near the epicenters of innovation. Those epicenters became what they are because intellectual, social, and economic capital are concentrated there, creating fertile ground for ideas to take root and scale. I saw this in all-hands meetings, in small group huddles, and in executive touchpoints. There was a constant pull and push: a bigger vision of the *who, what, and how* of moonshots, and a more contracted circle of the *who, what, and how*.

At times, the mythos is less compelling to me because it so rarely begins on the edge of life, where people are compelled to create out of boldness forged in the rawness of reality, not the well-traveled road of opportunity. It's not that traditional moonshot leaders lack this energy—many carry that fire—but their stories aren't always born from it. They're amplified because they're already standing where the world knows how to listen. (You might at this point be Googling exceptions; I'm not talking about them. I'm pointing to the pattern.)

No matter which project I worked with, there were moments when I could feel the room stretch around someone's words, a single idea seeming to fill the air. But the thing about big thinkers is that they're only ever as big as the world they can see. And the moonshot mythos depends on this paradox: it's built on both expansion and contraction. It expands the story of a few until their vision fills the entire stage, while it contracts the field of view, erasing the other possibilities that are always there. And it expands the potential to change the world through a moonshot, while contracting how we make that possible.

I recall a lunch with one of the architects of the early pipeline ideas for the next moonshot. He said something I'll never forget: "You know an idea, a moonshot, a moonshot thinker when you see them."

But what if what you see is a mirror and not an aperture?

It's not that these innovators aren't extraordinary—many are. But their scale and the ambition of the moonshot itself is shaped as much by the lens through which the world sees them as by their brilliance. Their names and ideas get repeated, their vision stretched to the horizon line. The spotlight expands. The world contracts. The myth needs both forces to survive: expansion gives the grandeur and possibility; contraction gives the power. It narrows, while giving a sense of expansion for what counts as innovation, where it happens, and whose stories are allowed to be seen.

The mythos is compelling because its scale is seductive. But it's also boring because it's a narrow vantage, dressed in grand language. It makes the future look like a private club. But the truth is that the future has always been crowded. Not everyone's light makes it through the aperture.

The myth, of course, isn't confined to people born inside epicenters of power. It reaches outward, folding in figures like Sundar Pichai. His story, growing up in a modest home in Chennai, India, is instructive. But when the aperture widens, it often absorbs the new figure into the same frame. The myth expands to make room for him, but it doesn't change its shape. I'd seen that move up close: the aperture shifting just enough to let someone in but never enough to redraw the outline itself.

So what does that mean for the rest of us? Perhaps the point isn't just who gets to innovate, but what happens to our sense of self and how we shape the world when we start to build our stories and imagination around theirs and existing moonshots. Entering the moonshot space means stepping into that dynamic. Sometimes you engage it; other times, it engages you. That, too, is part of the work: to recognize both what it gives and what it leaves out.

That recognition trailed me down every hallway at X.

Norm

It was an average Tuesday. I walked the halls of The Rails—the X building—past the micro-kitchen toward the café. Every few steps I'd hear a "Salima!" echo down the hall. Heads would turn. I'd smile, wave, and make a pit stop for a hug or a quick conversation. And then it would happen again, "Salima." Sometimes it was just to check on life. Often it was about a team, a project, a trajectory someone couldn't share with anyone else. Other times it was a quick huddle to move something forward.

That day my name echoed more than 20 times. It was the same ritual each time: people stopped, turned, smiled. I waved; energy shifted. By day's end, it struck me—after more than two years at Alphabet, I wasn't just part of the architecture of the moonshot system. I had become a presence inside it—a kind of frequency people gravitated to when they wanted to share unfiltered ideas, surface hidden challenges, or test strategies.

And under that current there was always an awareness: even here, in a place that prided itself on breaking norms, people were still pushing against invisible lines. Like Norm from *Cheers*, being a steady, recognizable energy blurred the edges. I reveled in it. I could feel that others did too.

Being inside the moonshot ecosystem is a little like standing backstage at a magic show. From the audience, it looks spectacular—effortless genius, world-changing vision, perfectly timed reveals. From the wings, you start to see the cues, the blocking, the lighting. You notice where the spotlight hits, what ideas echo and how it's cast.

When I first dropped into the ecosystem, it felt like a maze. On the surface, everything appeared open and fluid. What gave the maze its life were the invisible reference points everyone assumed you already knew—or could emulate quickly and flawlessly. To belong, you had to learn the language: understand the ambition of the companies, perform audacity, decode the unspoken rules of how to succeed.

My early days were overwhelming. For months, I carried a quiet trepidation. I knew I was standing in a place that bartered in technological possibility but pulsed with human dynamics. It wasn't just about what you could build; it was about how you learned to be inside the system. You were expected not to be constrained by the mundane realities of corporate life and still know how to play its game.

A chief human resources officer for one of the Bets once put the contradiction plainly: "You're here because you're smart. But you also need the right amount of curiosity, the right amount of authenticity, while playing the game of the system." That tension is one of the quiet architectures of the myth: genius environments cast as free zones of play and innovation but not fully.

The pages that follow are about the dynamics and conditions at the roots of Alphabet's moonshot ecosystem. There is an architecture that makes these projects possible. It's in grappling with these that we get at some of the moonshot secret sauce—the bit at the bottom that most people ignore but where the real flavors emerge. And grappling is what I did every day—the seesaw of values and ideas dancing with practice and reality.

The Test

The moonshot world didn't live in a single building, though X was its gravitational center. Some of Alphabet's future companies, Waymo, Wing, Intrinsic, took shape there before spinning out. Others built their own constellations elsewhere. What connected them wasn't geography but atmosphere: a culture designed to stretch what people thought was possible. The questions were audacious by design—*Do we build robots that make commercial production safer or tools that enable robots to do things not imagined before? Can self-driving technology change how people move and access opportunity? Is AI the new brain of our civilization?* The answers were built in fragments that filled whiteboards, labs, and hangars. To be part of that orbit was to learn how to shape something built from deep technical and business expertise—and aim it at problems the world had never solved.

That was the world I stood inside when I was called into Astro Teller's office one morning—a world where failure was engineered so technology could take its next leap and so could the people building it. I was excited to hear his thoughts and dig into strategy. For eight months, I'd been doing a systemic assessment of X—mapping the enablers and inhibitors for greater team, cultural, and mission-critical effectiveness. In short, where X could make its next leap in building a culture where everyone could succeed in launching the next billion-dollar moonshot.

I remember the moment that work became public. We were in an all-hands when an X'er asked, "Who's helping X get to the next level so everyone can do their best work here?" Sergey sat in the front row with Astro. Without hesitation, Astro pointed through the crowd like a dart and said, "Salima is working on that very thing. She's going to unlock that for us."

I was still jotting notes from the last question when I heard Astro's voice and felt 200 heads turn. My inner voice: *no, no, no, bro—you did not just land that plane on my lap*. The weight of X's cultural leap seemed to land there too. I sank 2 inches into my seat—at least it felt like that—then straightened up, smiled like a queen, and gave an exuberant wave.

"Happy to help."

So when I walked into Astro's office that morning, I was ready. Ready to hear his reflections; ready to sharpen the vision. We talked through the patterns I'd uncovered—the places where the organization was on the edge of something new. Then, as the conversation deepened, he shifted. His tone changed. The air between us stilled, as if something important was about to land.

"Salima," he said (paraphrased), "I have one question I ask everyone before they join X. If I told you that you could work on a project for six months with no job security, but it would be the most audacious project you've ever touched or you could work on something smaller, less risky, but with security, which would you choose?"

This was the moment. That quiet, charged pause when you already know what you're about to say.

"The latter," I said. "Not everyone has the life circumstances to choose the first."

He looked at me and said something like, "Then you failed the test."

A gut punch. A flash of doubt: *Can I be here? Am I not moonshot-worthy?* Then another thought: *But did I fail?*

In the moonshot world, failure is treated as one of its greatest teachers—a trampoline, not a trap. And what was happening wasn't a clash. It was a dance: two versions of audacity stepping into the same space and beginning to shape each other.

We both understood that risk matters, that boldness and the willingness to throw yourself into something bigger than yourself are the pulse of the moonshot space. You can't build something like Wing, a commercial drone delivery network with no well-worn path to market without that kind of boldness.

But audacity isn't the sole domain of those who can leap untethered. It also lives in people who carry ground realities with them, whose leaps are threaded with responsibilities, histories, and entire worlds they won't leave behind.

That was often my job at Alphabet: showing the Other Bets how to be relevant for more of the world. It wasn't about shrinking ambition. It was about expanding it. I remember chatting with Vindell Washington, an executive at Verily, Alphabet's biotech company: "What if your challenge isn't that you lack a good idea about making critical care more accessible—but that your version isn't shaped through the lens of the most vulnerable and underserved? What if you reimagined your customer? What if expanding the aperture of who you're building for is how you redefine the space of critical care?"

That work required its own form of audacity, not the leap away from the ground but the leap *with* it. Bringing people, realities, and histories into rooms where it's often assumed they don't belong. That's not soft. It's daring.

So no, Astro and I weren't at odds that day. We were, in a quiet way, building a version of audacity between us—one that held the exhilarating risk of flying into the unknown and the grounded truth of everything that travels with you when you do. That new version of audacity doesn't live in a vacuum. It needs something to push against.

That was my everyday work: walking into rooms and projects with the friction of real life at my back. It's not a well-worn path, but it's one you choose with clarity, knowing you're asking people—and technologies and businesses—to expand beyond the norm.

Audacity

Audacity is the unspoken currency of belonging in the moonshot space. I've seen brilliant people get hired and never quite find their way—not because they lacked vision or skill but because moonshots expect people to arrive already knowing how to be audacious in the way the system recognizes it. The space for orientation is narrow. If you're here, you're supposed to know how to shape a moonshot.

That assumption has its limits. If people were ushered in—shown where the power currents run, how the language works, how to push without vanishing, they could possibly build faster, bolder, better.

Audacity isn't a heroic trait you either have or don't. It's a practice: a willingness to be at odds with what is and to know when to align with what's in front of you. To step into the friction between being untethered and rooted. As my taiji teacher says, you have to be in the friction—push up against things, people, dynamics—to grow. Taking a moonshot from idea to thriving business is hard. Building breakthrough technology at the root is harder. In moonshot spaces, friction is constant. Some are trained by it. Others are pushed out. For those who stay, audacity isn't a spark—it's a compass forged in friction, guiding you through a future still being built.

This friction also appears in the building of the moonshot itself. A product leader once told me he wanted to push the variables and data shaping real-world use cases—so the technology entering the market would reflect how people actually live and work. But the incentives from earlier told a different story: shrink the research and development cycle to hit sales metrics, not real-world impact

It's easy to assume that if you're working in the moonshot space, you've already passed the test, that the moonshot itself will push out of mundane boundaries. But once you're in, another question emerges: *how do you and the moonshot live the audacity you've been recognized to have?*

Living Audacity

In the moonshot world, freedom is sacred—the belief that invention happens best when no one's looking over your shoulder. But structure is sacred, too—it's what keeps the chaos from collapsing on itself. The paradox is that you need both: freedom to imagine wildly and structure to make that imagination real.

The challenge for moonshotters who aren't in the highest echelons of leadership is learning how to locate themselves inside this tension. How much invention and play are valuable? How much do you color within the lines? A mid-level operations leader once told me it was hard to know how to shape something bold when you couldn't see the threshold—the line between being audacious and overstepping. Those who can navigate both—who can play at the edges without falling out of the frame—are rare.

I could feel this same force in rooms—the way people leapt into ideas and then pulled back, like invisible strings tugging them between boldness and caution. It wasn't physical, but you could feel it in the air—that charge of energy colliding with reality.

In one meeting about landing a new talent strategy, the tension was almost cartoonish. You could almost see steam rising from people's heads—ideas clashing, energy ricocheting across the room. Everyone was trying to hold the contradiction while quietly losing their minds over it.

The executives were frustrated. "There's too much process," one said, tapping a pen hard against the table. "We need people who can just jump in—start building, figure things out fast." Heads nodded. Another voice cut in: "Exactly. We need builders, not bureaucrats."

But then came the pivot: "We also can't have people going rogue."

A beat. Another voice: "Right, and we still need oversight."

The rhythm of agreement began to splinter. Voices layered over each other—urgency, exasperation, conviction. The energy in the room shifted from focused to frantic, as if everyone was arguing with the air.

And then came the clincher—loud, certain, almost triumphant:

> "We just need that entrepreneurial spirit—people who aren't worried about evaluations or set expectations!"

And yet, that desire—that fever for boundless, self-starting freedom—often came up against reality.

A senior leader had just joined one of the Bets, and she was told, "You figure out what your role is." For months, she wandered the halls—curious, observant, trying to find her orbit. To some, it looked like discovery; to others, drift. Her colleagues whispered, "She's doing nothing." Leadership said, "She's exploring." Colleagues said she was confused; leadership said she was tracking.

The truth was that she was building—but without any shared sense of direction, without expectations of what success looked like, the ground kept shifting beneath her. In our tea chat one day, she said quietly, "I feel like I'm free but trapped."

It stayed with me, that quiet contradiction, the exhaustion behind it. Eventually, she left. Not because she wasn't capable, but because too much freedom, without the frame of purpose, broke her. The projects she started never found traction.

I remember watching moments like these play out across the moonshot world. Rooms filled with brilliance and tension. It was like watching an orchestra where every instrument was perfectly tuned but playing a different song. Everyone wanted invention, but only within invisible boundaries. Everyone wanted freedom, but no one wanted the mess that came with it.

That's the paradox of the moonshot space: the same energy that drives possibility is constantly negotiating with the structures built to contain it.

You can feel this play out in people as much as in organizations. Some hold structure so tightly that they smother new ideas. Others hold freedom so loosely that nothing ever lands. The ones who thrive—the true moonshot builders—learn to embody both. They hold power lightly but deliberately, creating just enough shape for others to be brave inside it.

I remember speaking with Elliot Grant, CEO of a newly graduated project, Mineral, soon to become its own company, who believed transparency would rule. He learned that flatness isn't a shield against power—it just hides it. As CEO, he had to learn when to open the room and when to draw the line. The work wasn't authority itself, but accepting that authority was necessary.

Across the Bets, I saw this again and again. Leaders insisted their door was always open, that power was shared. In denying their authority, they didn't make the system flatter; they made power less visible, harder to navigate. That's another layer of the moonshot mythos: the story of power as shared, when in truth it's both shared and held, suspended between the desire for no structure and the need for it.

On the opposite spectrum, I saw a more mature Bet struggle to enable a fluid culture of learning and building. Its leadership clung to hierarchy, unable to recognize that space, not control, was the precondition for invention. Their desire for structure constrained creativity. The teams underneath could feel it: the ideas slowed, the risk-taking dulled, the spark that once animated the company began to flatten.

That's the real force that moves audacity. Not freedom alone. Not structure alone.

But the living rhythm between the two, the stretch and the root, the risk and the restraint that allows something impossible to take form.

Audacity doesn't live in a vacuum; it pushes against power, the structures that decide which risks get oxygen, which practices get room, and whose point of view becomes the sound of vision.

Living the Tensions

If you peel back the grand sense of what moonshots are, who they require, and how they come to be, all the talk of bold leaps and world-changing bets falls away. What remains is a quieter, sturdier architecture built on tension.

Every moonshot sits at the intersection of two competing forces: the need to imagine the impossible and the need to make it real. Those forces don't cancel each other out; they feed each other. The systems, rituals, and language of the moonshot world are all designed to keep that tension alive to stretch, but not snap.

At its best, that tension is what gives the work its brilliance. It's the hum in the building, the friction that keeps ideas awake. At its worst, it becomes a kind of gravity, the pull that can distort purpose, slow progress, and blur what the moonshot was meant to be in the first place.

To see the moonshot world clearly, you have to look at the scaffolding that holds it together: the belief systems, power dynamics, and operational myths that make it both extraordinary and challenging. You have to look at how culture and process, vision and hierarchy, audacity and accountability are constantly negotiating their boundaries.

Because in the end, moonshots don't fail from lack of ambition. They fail when the tension that fuels them goes unacknowledged, when the systems meant to hold possibility start holding it back.

When I looked across the moonshot ecosystem, I began to see that those tensions weren't random. They organized themselves into patterns—stories people carried about what it meant to work in a place like this. I came to think of them as the mythic tensions of the moonshot world: exceptionality and insularity.

Each one shapes how people see themselves in relation to the work, what they believe they owe the mission, and what they believe the mission owes them. Each one, in its own way, reveals the psychological architecture beneath the machinery of innovation.

Exceptionality

A colleague and I were chatting one afternoon at the micro kitchen, munching on almonds from one of those endless Alphabet snack jars, when he flipped his hair back and said almost casually, "Everyone here believes they were chosen." Just then, in the background, I heard someone else say, "No one is doing it the way we are." It was synchronicity. Like the myth had found a way to echo itself in real time.

That was the quiet expectation inside the moonshot world, specifically, that everyone here wasn't just brilliant but different from the ordinary world. It was the gravitational myth of innovation: that greatness requires separation. Inside these walls, people didn't just want to build something extraordinary; they wanted to *be* extraordinary. To be exceptional was to exist at a higher altitude, where the air felt thinner and the stakes higher.

I started noticing it everywhere. The way people walked, no matter which Bet they were part of. Like their feet were touching both the possible and the impossible at once. It was an unfamiliar feeling to me. Until one day it wasn't. I started walking that way too. A stride, a skip, a subtle glide, like the air itself carried purpose.

Exceptionalism was everywhere at Alphabet. It wasn't just encouraged; it was expected. You had to believe you were doing something no one else could do. That the impossible would break open precisely because you were the one standing at its edge. And to be fair, that belief was part of the fuel. You can't build a self-flying aircraft system or reimagine human intelligence without a certain self-centered gravity, a conviction that you can fly when others might only crawl.

But this belief didn't float in isolation. It anchored itself to something larger, the collective energy of being part of Alphabet and Google. To work there was to stand inside a world-shaping machine, an engine of global influence. The sense of exceptionalism wasn't just individual; it was symphonic. The mythos of innovation and impact became the shared oxygen everyone breathed. Think about how AI is—and will—reshape our reality—something Google is at the forefront of.

It makes sense in some ways. If you're working on a Bet like DeepMind, where machine intelligence is literally changing how humans experience their own, that's not just compelling, it's extraordinary. As Garry Kasparov once noted after observing AlphaZero's games, the machine's play was not only "surprising…and it also shows a lot of creativity," an assessment that many players felt reshaped their sense of what's possible on the board.[1]

I once introduced a world-leading scientist to one of the Other Bets. In his first conversation with a hiring manager, the very first thing he was told was: "We only hire the smartest people in the world. How do you compare?" He called me afterward and asked, "Is that arrogance or something else?"

It was actually a reflection of the mythos that lived in the walls, a conviction that to belong here was to be touched by something of great consequence. And the question always lingered, is this really true?

That conviction had power. It could animate brilliance, accelerate invention, and push the boundaries of what's known. But it also carried a psychological weight, a kind of invisible badge of worth that became hard to set down. Most people couldn't name it, but they lived it: the need to keep proving their exceptionality until they couldn't see themselves outside of it.

And yet, that same self-belief is also the fuel. You have to believe you can do something no one else can. You have to be a little self-centered to imagine a future that doesn't exist yet. That's the paradox at the heart of moonshots: the same impulse that can isolate you is also what lifts you into flight. Exceptionalism, at its best, is a kind of creative gravity. It pulls you toward the impossible, long enough to make it real.

Exceptionalism lifts and isolates in the same breath. It teaches people how to perform brilliance, how to narrate risk as courage, and how to turn audacity into a visible currency of worth. But it also builds a kind of glass wall separating the visionary from the world they might want to serve. The higher you climb into the myth of exceptionality, the harder it becomes to see the ground.

Insularity

All of us have incubated once in our lives, in our mothers' wombs. A cocoon, sealed off from the elements that could interfere with the larger biological purpose: birth.

Inside the moonshot ecosystem, it often felt like we were in that same cyclical process of seeding, growth, and birth or, sometimes, of premature endings. That rhythm wasn't only visible at the level of entire companies or projects but inside the micro-worlds of problems themselves.

When I'd talk with systems engineers at Waymo, they would often say that light detection and ranging (LiDAR) doesn't really have the same bias challenges as computer vision—that it simply measures distance, not people. And I'd ask: *but what about the data that makes its way into the LiDAR system?*

If we're building from the environmental and structural conditions of San Francisco, how does that account for roads, infrastructure, or foot traffic in places with different materials, densities, or layouts? If we don't include a variety of human forms and movement patterns, how can LiDAR ever learn to read them correctly? The sensor may not hold bias, but the world it learns from can still narrow its sightlines. And that narrowing isn't limited to technology; it mirrors the culture around it.

Alphabet was and still is its own ecosystem: vast, self-sustaining, endlessly resourced. Inside it, there is rarely a reason to look outward. The long-standing joke I came to appreciate over the years was that any time you asked someone at Alphabet for an outside connection or whether they had consulted anyone beyond the company, the answer was almost always *no*.

The world inside was so rich, so complete, that it became its own reference point. Expertise, validation, even critique could all be found within arm's reach. The risk wasn't arrogance; it was self-containment, a quiet belief that the answers were already here, waiting to be discovered by those on the inside.

I was having lunch with a colleague in my first few weeks at Alphabet. We were chatting about our life stories—what had brought us here, what I was looking forward to. Then she leaned slightly over her plate and, in a quiet voice, asked, "So…what's it like *out there*?"

I started laughing out loud, that full belly laugh. Until I looked at her and realized, *Oh shoot, she's being serious.* I quickly pulled back my amusement and shifted into a more serious exploration of the question. It was at that moment I realized how much this was a living and bordered ecosystem.

But that's also what made the place extraordinary. Incubation has its purpose. It's how new life or new ideas learn to breathe before they meet the air. Inside the cocoon, you're fed by what's immediately around you. You get to experiment, fail, and begin again without the noise of outside opinion or the premature pressure of the market. You learn the pulse of what you're building—the deep *why* beneath the *what.*

That kind of preservation can create astonishing breakthroughs. But over time, the same walls that protect can also distort. Insularity impedes expansion of perspective, of choice, of imagination. When the ecosystem becomes its own feedback loop, the external world starts to feel abstract, even irrelevant. And without friction from the outside, it's easy to mistake resonance for truth.

In practical terms, it showed up everywhere—in the way we used Google's own code libraries, relied on internal infrastructure, and took advantage of AI tools that were generations ahead of what the rest of the world could access. We could build fast, integrate seamlessly, and experiment with extraordinary precision. But it also meant that much of what we created couldn't automatically exist *outside* the Alphabet universe. The stack that empowers innovation also quietly reinforces the borders of where that innovation can travel.

It means that what's built *for* the outside world can, at times, be *outside of the outside world.*

Every incubation eventually meets its test: the moment it must face air and gravity. And when it does, it often performs beautifully in places of similarity, environments that mirror the conditions it was born in. But the real test comes in places of difference.

Could a Waymo self-driving car, trained on the ordered chaos of Mountain View, survive in Mumbai? But what if it were built *as if it could from the start*? What new forms of intelligence, adaptability, and resilience might emerge if innovation were designed to thrive across difference as a necessity, not as an add on or retreat from it?

I remember being in a meeting with the leadership of Loon—a now-failed moonshot bet. We started talking about Kenya, one of their test markets. I asked, "Have any local engineers or partners weighed in on how our solution could work in their environment?" Then I paused and said, "Actually, let me take a step back—are the Loon teams aware of Kenya's long history with foreign entities arriving to present innovation solutions not built *with* Kenyans?"

The room went quiet. Not defensive, just still. It wasn't malice or oversight, just the quiet inevitability of a system that had learned to innovate in isolation.

As I was walking out, one of the senior business leaders said, "But Salima, what's relevant for us is whether our technology *works* there." And I replied, "Yes, but *how* and *for whom* it works matters perhaps even more."

It's easy to see insularity only as a limitation, but in the moonshot world, it served a twofold purpose. It was incubation, a practice of seeding, growing, and birthing ideas inside a self-sustaining environment. And it was preservation, not of fragility, but of exceptionality. A way to protect the conditions that make the impossible feel ordinary, that rare atmosphere where ambition can expand without friction.

To be honest, it was exhilarating. There's nothing quite like being in a place where you're not constrained by money or imagination. For me, it made that little girl, the one who once held a wad of cash in her hands, imagining what she could invent if resources were real, suddenly feel seen.

But it was also constraining at times. Reaching out to other industries, communities, or even ideas beyond our walls often felt like an exception rather than a fundamental part of the practice of moonshot building. And there were really no incentives to look outside. In conversations with project leaders or product teams, it often seemed like reaching beyond the Alphabet ecosystem was viewed as an unnecessary burden rather than a valuable addition.

It was almost as if too much was available inside—no gaps could reveal themselves. And if you did look outside, it felt like you didn't get the groove. You were out of step. But what if you got the groove but just knew there were more rhythms, more ways of working and thinking that could have made the groove even better.

The paradox is this: insularity doesn't include the outside world until the very last mile. By then, we've already built things that have to *fit into* the world, rather than *with* it. One of the Bets even graduated into a business without a real market fit or opportunity strategy. It was a beautiful idea but one without a world to belong to.

I think of my father serving customers behind the counter. I can still hear the ding of the register. I would sit on a milk carton, eat ketchup chips (very Canadian), and drink Coke, watching him move through the day, how he, in his way, created out of what was available to him.

He didn't call it innovation. He called it "dhandho" (business). But even then, I could see that he was building something larger, a rhythm between what people needed and what he could provide. His imagination lived inside limits, not beyond them.

Maybe that's why I've always been drawn to the spaces where constraint meets possibility, where invention isn't about abundance, but about attunement. What I learned behind that counter was that creation begins not from having everything, but from paying attention to what's already in your hands and in the hands of others. And what I learned at Alphabet is that the extraordinary is more than possible, not through ease but through friction.

The dynamics and conditions that exist within the Alphabet moonshot space are their own. I'm sure you'll find parallels and overlaps elsewhere. But the point isn't to re-create the Alphabet moonshot ecosystem. It's to use its lessons as a trampoline for what else is possible—to create a moonshot that can hold any internal world, with all its resources, dancing with what's external, with all its resources.

Because that's the next frontier—expanding what a moonshot is, who builds it, and how it changes the world.

Now, let's 100X the 10X moonshot.

CHAPTER

3

The Third-Space Moonshot

The Signal and the Instinct

I want you to go somewhere quiet. Somewhere no one will interrupt you.

Close your eyes and take seven slow breaths in and out. Now imagine a moment you wanted to build something—whether you believed you could or couldn't. Where did that instinct come from? What was the signal you were responding to?

Every moonshot begins with an instinct. Not the loud, cinematic kind we like to attribute to genius, but the quiet internal pull toward something that must be built. We romanticize moonshot ambition as vision, brilliance, courage. But before any of that, there is something more primal: the impulse inside you to create with the world. Something that changes it, expands it, or reorders it.

Beavers don't set out to engineer ecosystems. They don't gather for strategy off-sites or draft linear roadmaps. They hear running water and feel compelled to build a dam. Play a recording of trickling water from a speaker, and a beaver will immediately begin constructing a dam over the speaker. It cannot help itself. The instinct to build and protect is its orientation to the world.

And yet, that simple, nearly comical compulsion—*water is moving; I must stop it*—ends up unlocking entire ecosystems: wetlands, biodiversity hubs, flood mitigation systems, water filtration networks, and new microclimates. But before any of that happens, the beaver must hear the signal. It must trust it. And it must move toward it.

This chapter is the beginning of a practice for shaping what we build once we follow that inner signal—whether cultivated or innate.

The Architecture of Moonshots as We've Known Them

What we've historically called moonshots, especially inside Silicon Valley and Alphabet, has been built on four intertwined layers. Each one is powerful. Each one has its limit. Together, they create a paradox: the very conditions that enable extraordinary breakthroughs also constrain how far those breakthroughs can reach.

- **The Solution Layer—what we built.** Highly audacious point solutions aimed at massive problems: self-driving cars, stratospheric connectivity, robotics, identity, sensing, health diagnostics. This is the visible invention—the part the world recognizes as a "moonshot."

- **The Incubation Layer—how we built it.** A uniquely resourced environment: deep capital runways, elite technical talent, protected time, and political air cover. Freedom to fail. These inputs made the improbable feel inevitable.
- **The Cultural Layer—what shaped it.** The myths, norms, and emotional architectures that guided who thrived and why: exceptionalism, speed as virtue, audacity as currency, and a gravitational pull toward "brilliance."
- **The Problem–Technology Boomerang—the dynamic powering it.** A constant oscillation between a compelling problem and a seductive technological possibility, each reinforcing the other. This boomerang fuels discovery, but it also narrows the aperture of what counts as a moonshot.

The Paradox of the Lab

Underneath this architecture lie the traditional inhibitors:

- **Long time to impact:** 10–15 years on average to bring a moonshot into the world
- **Massive capital requirements:** Billions invested, with a low survival rate for early concepts
- **Diminished value:** Innovation silos that abstract inventions from the very ecosystems they are meant to serve

Despite these constraints, this architecture has produced stunning technologies that have impacted millions. Yet it has also kept moonshots incubated *outside* the world—away from the partners, markets, social intelligence, and inflection points of possibility that could have made them even more impactful.

This is the paradox: the lab protects the spark, but it can also starve the flame.

The Threshold We've Reached

The Shift from 10X Products to 1000X Systems

To move into the next era—the 10X to 10X, the 100X, the 1000X—we cannot simply refine the old model. We must expand it. Even build anew. Not because the Silicon Valley lineage is wrong or obsolete but because it represents one set of inputs in a world now demanding a far larger repertoire.

There will always be point problems requiring point solutions. Thousands of startups will (and should) keep doing that. But when we talk about building solutions that benefit not just millions but *hundreds* of millions and shift entire industries, societies, and planetary systems, we need a different ambition, a different orientation, a different set of capacities.

My hypothesis is this: *we have been too slow, too insulated, and too safe*. The world now requires a new kind of audacity. One that is engaged with reality, responsive to multiple contexts, and capable of moving whole systems, not just products.

This "third-space audacity" opens the door to moonshots that are cross-regional, adaptive, revenue-generating at new scales, and net-positive for societies. These are moonshots that few have imagined and even fewer have known how to build—moonshots where technology becomes a co-pilot to unlock flourishing, ingenuity, and compounding forms of value.

To entertain such possibilities is both exhilarating and unsettling because this orientation is bolder than bold. But like any worthy moonshot, we are here to expand the very definition of what deserves a moonshot and how it can come into being.

But to be clear, not every need, opportunity, or problem becomes a moonshot. Nor should it. The point is not to force a moonshot onto everything. The point is to recognize when a moment, a system, or an opportunity calls for it—and to be willing to dive in and do the unprecedented. And there are contexts where no solution or technology is required at all. Not every need calls for intervention, and not every human or other system needs to be reshaped.

Expanding the Inputs of Moonshot Creation

For several decades, one lineage dominated: speed, technical audacity, concentrated talent, and the belief that breakthrough technology could do for the world what hasn't been done before. It set the standard for ambition. But it never captured the full landscape of how transformative solutions emerge or how value and markets can be unlocked at scale.

There has always been a wider repertoire: ecosystem awareness, contextual intelligence, cultural imagination, temporal depth, ingenuity under constraint, generational pattern-recognition, past technological innovations. These forms of knowing were not always absent. They simply weren't integral inside the dominant architecture of moonshot buildings.

We are now entering a world that requires a larger field of view, not to satisfy a moral ideal but to expand our capacity to build what this moment (and future moments) will demand. As we widen the aperture, new forms of value come into reach: new markets, new pathways for capital, new engines for coordination, and new ways technology can serve more people, industries, and regions.

This is not about replacing one lineage with another. It is about building a more capable, more ambitious repertoire for moonshot creation; it is about an expansion of possibility, not a rejection of one inheritance.

With that orientation, what I'm proposing here isn't meant to close the door on other ways moonshots can emerge. The very ethos of a moonshot is to push beyond what we think is possible, which means there will always be forms we haven't imagined yet. Some may not reach millions. Imagine, for example, finally curing a rare disease that has plagued a small community for generations. The numbers may be small, but the breakthrough transforms lives that had long been written off as intractable. Scale alone is not a marker of a moonshot. What matters is the willingness to challenge the boundaries of what we believe can be solved—and to mobilize the people, ideas, and technologies needed to make it real. My call for third-space moonshots is not meant to narrow the field of ambition, but to act as a slingshot into even greater possibility.

Mapping the Two Repertoires

Dimension	Traditional Lineage	The Expanded Repertoire
The Moonshot	An audacious problem and a breakthrough solution	Ecosystem-level solutions that meet global needs and impact hundreds of millions
The Driver	A new technology frontier	Technology as enabler, amplifier, and integrator
Capability	Concentrated mastery, elite specialization	Cross-disciplinary, cross-cultural, and cross-regional intelligence
Value	Market valuation, speed, and scale	New revenue streams, new economics, and new markets unlocked
Impact	Solving a massive problem for millions	Compounding effects across interconnected systems
Success	Billions in revenue	Unlocking new ways of living, participating, and thriving
Time	Long arcs of deep tech development	Adaptive solutions and opportunities that can outpace the problem
Blind Spots	Relational, contextual, and systemic dynamics underutilized	Surfaced, integrated, and converted into value

Silicon Valley's Temporal Instinct and the Intelligence of the Third Space

One of the most powerful inputs that shaped the traditional lineage of moonshots is its instinct for time: *move fast, learn fast, scale fast*. It is a temporal stance—a way of interpreting urgency, momentum, and progress. But it is only one orientation toward time.

Across much of the world, innovation follows a different temporal pattern: one shaped by relationships, seasons, communal rhythms, and the long arcs of continuity. Not slower. Not faster. Simply attuned to different signals. In the third space of moonshot building, neither rhythm is sufficient on its own:

- **Speed without attunement** outruns the worlds it hopes to serve, creating products that arrive as "foreign objects" rather than landed solutions.
- **Attunement without momentum** generates profound insight that never reaches the scale its wisdom deserves.

The third space demands a Temporal Intelligence that blends velocity with sensitivity. It is moonshot time reimagined: time that expands or contracts as needed, recognizing that the "first mile" of trust often dictates the "last mile" of scale. When we align our breakthroughs with the ecosystems they are meant to impact, we find that adaptive solutions can actually outpace the problems they were designed to solve.

The Shift to Systemic Orientation

When I sat with Other Bets CEOs and asked, "How do you want to arrive at achieving your mission?" I was offering a worldview shift. I was asking them to acknowledge a deeper truth: any moonshot worth building is, by definition, an ecosystem.

You cannot build a moonshot without building the environment that will sustain it. When leaders stop seeing their mission as an isolated intervention and start noticing the forces shaping it, the matrix comes into view. They stop optimizing for narrow metrics and start optimizing for cascading impact.

Moonshots arise at the intersection of grounded realities and enabling technology. Not tech for tech's sake. Not idealism without infrastructure. But in the third space where ecosystem-level change becomes possible.

This is the Four-Foundation Practice for seeing the full field.

The Third-Space Moonshot Practice

Collective Orientation: Seeing the Field, Not the Fragment

When most people want to identify a moonshot problem, their instinct is to start researching. They scan trends, talk to experts, and read reports. They tap into something they are personally passionate about. But that still keeps you in the fragments.

A third-space moonshot begins somewhere else entirely, because we are looking to solve something unseen, undervalued or hidden—something in an ecosystem waiting to be unhinged. We begin by bringing together the full field of capability because we are trying to discover what might exist—not simply find what we already know how to look for. Not one mind searching, but many minds shaping what becomes visible. Scientific insight, lived experience, cultural knowledge, local expertise, institutional memory, and creative instinct all held in the same space, at the same time.

Ecosystem moonshots rarely originate from a single vantage point. They arise at the moment these perspectives collide, overlap, and illuminate each other. The starting question shifts from "What can I solve?" to:

> "What becomes visible when the full field of intelligence gathers to sense the problems, needs, and opportunities ready to be engaged?"

This reframes collective creativity and lived experience as power positions. Because when that intelligence comes together, the ecosystem doesn't just reveal needs; it reveals latent value. It uncovers the different logics through which we identify challenges, a broken part of the system, the untapped markets hiding in overlooked communities, and the economic flows that could be redirected or amplified.

Think about how often a problem changes shape the moment another perspective enters the room. A colleague notices a constraint you had taken for granted. A community member reframes the issue entirely. What first appeared as a technical challenge suddenly reveals behavioral dimensions you hadn't considered. What you thought was now looks different because we made a discovery together.

In the third space, we start with a collective orientation:

- Orientation that challenges, expands, contradicts, and surprises
- Orientation that is skeptical and hopeful at the same time
- Orientation that is deliberately situated in different contexts, disciplines, and lived realities

This is when the ecosystem starts to reveal itself. This is a multilayered collective compass—the moment when the real map of the ecosystem begins to emerge. Momentum amplifies here. Not through individual speed, but through collective acceleration. This is how we become more sophisticated in the "first mile" of finding the moonshot: not by searching harder, but by seeing more, with greater clarity and layered perspective.

In practice, this means bringing creatives, operators, technologists, workers, funders, policymakers, and people living inside the system into a process of owning the discovery of opportunities, solutions, and needs. But this requires a shift in incentives. If your goal is simply to identify a problem and deliver a solution, you will follow a very different path than if your goal is to discover a moonshot need—something that becomes visible only when many parts of the ecosystem are in the discovery together.

Strategic Complexity: Designing Within, Not Around

I once asked someone from a health tech company a simple question: "Have you ever missed a doctor's appointment?"

They had been showing me their appointment tracking and scheduling app—smart reminders, automated follow-ups, elegant interfaces designed to make it easier for patients to keep their appointments. And yet the data still showed the same thing: people were missing appointments.

As we talked, another possibility surfaced. What if people weren't missing appointments because they forgot? What if they were missing them because their lives were complicated—because missing work meant losing wages, because transportation wasn't always accessible, because childcare fell through?

The technology assumed the problem was forgetting. But forgetting was only the surface of something much larger.

If there is one instinct that has shaped traditional technology building, it is this: simplify the problem, strip away the context, isolate the variables, and then build the solution. This approach makes products possible. It makes engineering tractable. But it has also lifted solutions out of the very worlds they were meant to transform. When you simplify too fast—when you pull a problem or solution out of the living complexity of people, relationships, incentives, misalignments, and the quiet opportunities that breathe between them—you strip away the very intelligence that reveals what's truly possible.

In the third space, our instinct is different. We don't step back from complexity. We step into it.

We don't enter the field to drown in it or to admire it; we enter it to let complexity reveal what only it can show. We enter the living, breathing field of the humans, the frictions, the networks, and the constraints that feel immovable—until suddenly, they aren't. It is in that field that what needs to be built becomes visible.

And at that point, something shifts. An opening appears that wasn't visible from the outside. A small action feels capable of reorganizing an entire system. A part of the ecosystem signals that it is ready for change. Technology stops being the hero and becomes the accelerant, the amplifier, the partner.

This is designing in complexity, not around it. It is what allows us to identify the places where a single, well-placed intervention can create ripple effects that touch millions—and eventually hundreds of millions. This is where moonshots stop being isolated solutions and start becoming ecosystem unlocks.

Complexity is not the enemy of clarity. It is the source of it. When you stay with the complexity long enough, something unknown emerges—like lightning seen for the first time. A new opening. A pattern that wants to reconfigure itself. A shift that changes the scale of what's possible.

This is the instinct the next era of moonshots requires: the courage to enter complexity, the sensitivity to read it, and the audacity to move through it with precision.

It's not softer. It's not slower. It's not "nice."

It is more powerful because it lets us design solutions rooted in the world as it actually is so we can build the world that is ready to emerge.

Anatomy of an Unlock: GoJek

Evidence of the Informal Economy Activated Through Systemic Coordination To understand these first two foundations in motion—seeing the full field and acting precisely within complexity—we turn to GoJek.

GoJek emerged not as a simple transportation app but as a coordinated intervention into Indonesia's informal economy—one of the largest and most intricate informal systems in the world. Its insight was in seeing the underlying field: millions of motorcycle taxis (*ojeks*), fragmented micro-merchant networks, and a population largely excluded from digital payments. By placing technology at the system's highest-friction nodes—real-time mobility, merchant logistics, and cash-dominant transactions—GoJek activated an economic circulatory system that already existed but had no connective tissue.[1]

As founder Nadiem Makarim has pointed out, GoJek was about improving livelihoods at scale. By 2020, GoJek contributed more than $7 billion USD to Indonesia's economy and expanded income pathways for millions of drivers and small merchants.[2] The platform is a mechanism for economic participation, highlighting how the system's coordination created forms of value that no single actor could have unlocked alone.

But the real insight for the third space isn't just in the success; it's in the aliveness of the system that followed.

As with any ecosystem-scale intervention, new dynamics surfaced—the kind that naturally arise when a dormant system is activated. Driver incomes became intertwined with platform rhythms; merchants gained digital visibility while adapting to algorithmic marketplaces; and regulators encountered an ecosystem evolving faster than governance structures. These are not shortcomings; they are signals. They reveal the interdependencies that awaken when technology reorganizes economic and social relationships.

Viewed through a third-space moonshot lens, GoJek is a living system that is still evolving, pointing toward additional horizons of possibility. We have to grapple with what comes next:

- What new forms of shared digital infrastructure could deepen financial inclusion beyond a single platform?
- What models of value distribution or co-governance could enhance long-term ecosystem resilience?
- Which emerging leverage points—local entrepreneurship networks or community logistics hubs—might open the next order of economic participation?

Rather than a completed moonshot, GoJek offers a way of seeing: how an ecosystem reveals its design evolutions once its initial unlock has occurred, and how the next 10X emerges from reading those signals.

Technology as an Enabler: Placing the Instrument Where It Can Sing

In the third space, technology is never the hero. It is an instrument—the means through which we create possibility, enable a solution that was previously unthinkable, and take human behavior and experience into new arenas.

In the traditional moonshot, you are incentivized to build technology wrapped in a sci-fi mythology—something intentionally, almost performatively, hard to achieve. But in the third space, novelty is not the goal. The goal is to make technology do what it is meant to do: create ecosystem-level transformation. This could mean building a new technology altogether, but it could also mean using previously discarded tools and finding new applications for them. It could mean using fragments of existing tech in a new way. Whatever form it takes, the key is that the technology is unlocking audacious 10X to 10X opportunities by acting as the lever that:

- Brings new economic value across an ecosystem
- Solves more than one need or problem at once
- Brings people, communities, and societies into another way of living and well-being

If this is what technology must enable, then we have to be strategically precise about where we place it. The work demands that the "instrument" is placed in the highest value node—aligned with an opening to readiness and connected to real relationships. This is how we elevate human and planetary life.

At a more everyday level, think about it this way. My four-year-old son was experimenting with an electric soap dispenser the other day and began asking:

> "What happens if we put it by the faucet?"
> "What happens if we mount it on the wall?"
> "What happens if we put it under the sink?"

What he was really exploring was the significance of placement. Where you position the instrument changes what it can do, how you experience it, and what it makes possible.

But this placement is never static. The technology, the opportunity, and the effect must grow, shift, and move to remain useful. In the third space, we don't just launch and leave. We stay in the friction to ensure that potential harms or unwanted impacts are out-designed at every stage of the evolution—and that value and benefits are compounding. We are not just building for impact; we are building for the sanctity of the journey.

To find that node, we have to ask questions like:

- Where has technology worked in this ecosystem in the past, and what are the lessons?
- What has failed?
- What are the inhibitors and enablers to innovation, and where are the openings?
- By placing tech here, what are the upstream and downstream implications?
- Who benefits? Who loses? Whose lives change?
- What transitions are required?
- What social, cultural, and economic forces does the technology need in order to be received, integrated, and operationalized?

The third space does not reject the past, our repertoire is limitless. We can draw from AI and machine learning, robotics, and satellite constellations, but we can just as easily draw from shortwave radio, vacuum tubes, or quantum technology. Nothing is excluded. The real questions are: Where does technology enable a solution that is trying to emerge? How does its placement enable ripple effects of value, impact, and outcomes?

Moonshot technology is transformative when it changes micro and macro realities, enables a new way of being, and becomes a platform for many rather than a product for a few. The power is never in the novelty; it is in the precision—the way it acts as the connective tissue that allows an ecosystem to reorganize around a new possibility.

Expanded Outcomes: Building for the Future We Actually Want

Traditional moonshots are trained to look for a narrow kind of proof: a breakthrough technology, a billion-dollar valuation, or millions of users. Those markers matter, but they tell only a fraction of the story.

A third-space moonshot widens the aperture. It recognizes that when an ecosystem begins to shift, value does not appear in one place—it ripples. It shows up across people, institutions, markets, and mindsets. This widening is not philosophical; it is practical. It is how more futures become possible at once. In this expanded frame, outcomes look different:

- **They look like** new income streams flowing through multiple parts of a system, or many systems for many actors.
- **They look like** markets being nurtured into existence, not extracted once they're visible.
- **They look like** value rediscovered in places long labeled "undervalued"—communities, capabilities, and ideas that were always there, just unseen.

- **They sound like** a spirit of invention spreading outward, where creativity becomes participatory rather than concentrated—something people step into, not wait to be invited to.
- **They feel like** billions in latent value unlocked, not through hype but through connection, when the ecosystem can finally see itself and coordinate.
- **They behave like** systems that grow stronger as they scale, sometimes even regeneratively, or new systems are produced all together, where each layer creates the conditions for the next to thrive.

And at their best, these moonshots create alignment: shared problems named clearly, shared needs acknowledged honestly, and shared opportunities acted on together. Even if it does not happen all at once, different actors invest in the effort and experience its benefits at different points across the ecosystem.

Consider the shared outcome of ensuring long-term income opportunities in the age of AI. Many actors may align around that aim, but they will not experience the benefits at the same moment. Young entrepreneurs may see the upside first as new ventures and capabilities emerge. Governments may see it later as solutions move through public education systems and workforce programs. Labor markets may feel the shift later still, when the transformation of work becomes not just displacement but substantive gains for workers.

This is what makes coordinated moonshot building possible at scale.

These are not side benefits. They are not "nice to haves." They are the signature of a moonshot built with attunement and collective intelligence.

Estonia's X-Road offers a glimpse of what becomes possible when infrastructure, governance, and technology move not in isolation but as a coherent system—each reinforcing the other, each expanding what the future can hold.

Anatomy of an Unlock: Estonia's X-Road

Evidence of the Administrative Moonshot and the Shared Digital Backbone Estonia's X-Road began as a quiet administrative innovation and became one of the most consequential digital infrastructure moonshots of the modern era. After regaining independence in 1991, Estonia faced a radical challenge: rebuilding a state from the ground up with almost no resources.

Instead of digitizing ministries one by one, the country recognized a deeper truth: government, banking, healthcare, and education all required a shared, secure, interoperable backbone. The insight was that the state itself functions as an ecosystem. Digital identity and data exchange needed to become **infrastructure**, not scattered services. X-Road—the secure, distributed data-exchange layer—emerged from this principle. Estonia was not simply building digital tools; it was building the **connective tissue** of a digital society.

The outcomes of this unlock are measurable across every sector. Today, more than 99% of public services are digitized.[3] According to a cross-country assessment of digital public infrastructure, Estonia's e-ID and X-Road platform save the equivalent of about 2% of GDP annually through administrative efficiency.[4]

These efficiencies compound: a thriving fintech ecosystem and seamless public–private coordination all trace back to this digital backbone. As Estonia's leadership describes it, the government operates as a platform—a system designed for transparency and shared value.

But the third space asks us to look closer at the dynamics that emerge when a nation becomes interoperable at this depth.

Trust becomes a shared asset rather than the responsibility of a single institution. Public–private boundaries blur as banks and hospitals become co-producers of the same infrastructure. These are not complications—they are the signals of an ecosystem coming alive.

Yet, we must grapple with the next horizon:

- What might it look like for citizens to dynamically control data flows—expanding consent beyond static checkboxes?
- How might regional digital identity systems evolve into a cross-border "digital commons"?
- What forms of shared cybersecurity or federated trust are required as societies become even more interconnected?

X-Road is not a finished moonshot. It is a platform for future moonshots, revealing how infrastructure and relational intelligence co-evolve to unlock entirely new societal possibilities.

The Essence

A third-space moonshot is not built from ideology, binaries, or inherited playbooks. It arises from attunement—to the ecosystem, to emergence, to timing, and to the many forms of knowledge and experience that exist in the world.

It asks a driving question:

> What wants to emerge here, and how do we meet it with our collective capability, creativity, and strategic imagination?

This is the position of power. This is the practice.

When all four foundations converge—collective intelligence, interconnected complexity, technology placed with strategic precision, and outcomes that strengthen the whole—moonshots take on a different quality. They become generative systems rather than singular solutions. This is where the third space lives: in the expanded horizon of what becomes possible when invention is attuned to the full system it is entering and the new architectures it is creating.

Planet Labs stands at this edge. It opens a new aperture on the world, revealing how perception itself can become a catalyst for the next order of breakthrough.

Anatomy of an Unlock: Planet Labs

Evidence of the Planetary Nervous System and the Shift from "Battleships" to "Smartphones" Imagine being able to see the entire Earth every single day, not as a static photograph but as a living pulse. Coastlines shifting, forests thinning or thickening, crops gaining or losing color, floods rising, coral bleaching, ships gathering in unexpected places. A planet in motion. When I first encountered Planet Labs' work, this was the feeling: not a product, not a dataset, but an aperture opening into a different order of perception.

Planet Labs did not begin with the question, "How do we build a better satellite?" Reflecting on his time at NASA, Chris Boshuizen described how space programs had been building satellites "like battleships" and asked what might be possible if they were built more like smartphones instead.[5] This wasn't just about size or cost; it was a reframing of purpose. Instead of building rare, exquisite machines intended for narrow use, they imagined a constellation of small satellites creating a continuous record of Earth.[6]

In my own experience working with moonshot teams, I've learned that this shift in intention—moving from building rare, exquisite machines to building a continuous, shared vantage point—often signals the beginning of a true moonshot. It's not about novelty; it's about repositioning technology inside a field that is ready to reorganize.

The founders realized that Earth scientists, conservationists, farmers, disaster responders, NGOs, and governments were all operating with partial visibility. The world was changing faster than the tools designed to monitor it. Behind the mission to "use space to help life on Earth" is a third-space logic: the recognition that when systems evolve quickly, the greatest unlock is not solving a single problem, it is improving our collective ability to perceive change itself.

Planet Lab's satellites—lightweight, inexpensive, launched in clusters—created temporal abundance. Not the sharpest image, but the most frequent. Frequency reveals pattern. Frequency reveals stress and resilience. As Will Marshall, CEO of Planet Labs, has often argued, if you can see change, you can manage it.[7]

As imagery began refreshing daily, the ecosystem around it shifted. Humanitarian teams used it to coordinate emergency response. Agricultural companies monitored crop health in near–real time, detecting early signs of stress. Climate researchers tracked deforestation with a precision that changed policy and enforcement. Insurers, supply chains, and financial institutions embedded this new layer of Earth awareness into their risk models. Robbie Schingler described this effect as creating a "common operating picture" of global change.[8] This wasn't a transformation of a single industry; it was an activation of the interdependencies between them.

What makes Planet Labs compelling is not just what it built, but the horizons the system reveals next. Daily Earth observation has already transformed fields far beyond the environment. Humanitarian teams detect early signs of displacement; infrastructure managers track roads, grids, and pipelines; supply chains verify the movement of goods; investors evaluate ESG claims against observable reality; and governments monitor everything from groundwater depletion to illegal fishing. The economic implications are significant: markets become more transparent, risk premiums adjust faster, and new financial instruments emerge from verified planetary data.

One of the principles I've learned from years inside moonshot environments is that these types of tensions—where high-frequency imagery meets traditional policy or where digital transparency meets established supply chains—are often signals. They are indicators of where the next design horizon lies. They reveal the aliveness of the system.

As we look through the third-space lens, the grapple continues into new horizons:

- **The integration horizon:** How does high-frequency imagery merge with radar, thermal, acoustic, and ocean-based sensors to build a fuller planetary intelligence?

- **The decision horizon:** How could urban planners anticipate infrastructure stress before outages or food systems stabilize through shared visibility into soil and water cycles?
- **The sovereignty horizon:** How do local communities and Indigenous nations co-govern parts of this intelligence, creating new forms of data sovereignty?

The satellites were never the end point; they were the beginning of a relational system where business, government, community, and ecology move with a shared, verifiable understanding of the world. The moonshot is the emergence of a planetary nervous system—distributed, participatory, and capable of helping humanity perceive the Earth with enough clarity, rhythm, and relational intelligence to act differently.

The Opening

So, let's bring this back to where we began.

The Silicon Valley–Alphabet lineage of moonshots is one path—powerful, proven, and resource-intensive. What you've encountered here is not a replacement. It's a leap. You are not being asked to choose; you are being asked to imagine differently.

Most people in the world will never have access to millions in runway or tightly protected innovation labs, and that is precisely why the next leap in moonshot building matters.

The task now is to draw on lessons, methods, and orientations from many places and build from the third space—where technology becomes a partner in expanding human and planetary life, and where bold outcomes are no longer the domain of a few, but emerge through strategic, audacious ecosystem-level intelligence in action.

Once you see this path, moonshots stop feeling distant. They are ready to be found and built.

Moonshot Dojo

This is where the practice happens.

If you were to walk out of this chapter and try to build something bold, certain thoughts would likely surface immediately.

Good.

Now we spar.

Reflex: Simplify the problem. "This is too complex. Let's narrow it down."

Translation: Too much. Make it stop.

Dojo move: Stay with the complexity and let it unravel you.

Ask: What disappears when we simplify too quickly? What appears when we don't?

Reflex: Stay in your lane. "Why would I build with people outside my field or known world?"

Translation: Too many cooks in the kitchen. Some don't even know how to cook!

Dojo move: Expand the field.

Ask: What will we miss if we limit who and what shapes this moonshot?

Reflex: Build the solution quickly. "Let's just create the technology/product, do our basic due diligence, implement it, and the rest will adapt."

Translation: I *know* what you need.

Dojo move: Look. Listen. Engage. Then move.

Ask: What could this technology unlock here—and where must it be placed for that to happen?

Reflex: Move faster. "This will take too long."

Translation: I feel jittery. Go faster. We'll lose out. Faster.

Dojo move: Flow in the tension.

Ask: What signal might emerge if I stay in the mess long enough to see it?

Moonshots rarely fail because people lack intelligence, talent, or technology.
We limit what could be when we collapse the field too quickly.
Notice the reflex.
Spar.
Let the moonshot come alive.
Then build from there.
Outcreate what shrinks possibility.

CHAPTER

4

Moonshot Leadership Begins the Moment You Say Yes

The Mirror to the Moonshot

There I was, with 10 faces staring back at me. The energy in the room was completely awkward—like 10 dudes arriving at a school dance and standing there, not knowing how to interact with each other. These leaders had never been on the same call together.

Ten CEOs of Alphabet's most ambitious moonshot bets—people who ran companies like DeepMind, Wing, Google Fiber, Sidewalk Labs, CapitalG, and Waymo. A mix of Silicon Valley veterans and leaders from telecom, government, and traditional corporate industries. An unexpected cocktail of backgrounds, expertise, and egos. Some knew each other; others knew of each other only in passing.

I had invited them to a meeting on how we might more strategically integrate their companies, missions, and technologies with undervalued markets, users, and talent. I genuinely didn't know who—if anyone—would show up. So when almost every CEO said yes, it told me something: the moment was right.

But that "yes" didn't happen in a vacuum. It was the result of years spent in the trenches with each of them individually. Over time, they had come to appreciate my role as the person who expands perspective, mission, and action—not as a philosophical exercise but in ways that added direct value to their business and moonshot goals. I had become a trusted navigator, someone who could help them see the opportunities they were missing without losing sight of the hard-edged metrics they had to hit. They knew I wasn't there to add "fluff"; I was there to add dimension.

The conversation ignited quickly. Could this be done? What kinds of incentives or accountability would make it real? How might they, not just as technologists but as leaders, be valued for building with contexts and realities Silicon Valley rarely centers?

Debate ricocheted around the room, ideas clashing and reforming. And then, toward the end, David Crane, CEO of GV, spoke up with disarming clarity.

"I nominate Salima to bring this to Sundar. We trust you. Anyone who agrees, raise your hand."

Every hand went up.

Mic drop. Me? Right. Right! Voluntold. March upstream and deliver that vision on their behalf. No pressure. What struck me wasn't just the nomination; it was the dynamic underneath it. In that room, something subtle but significant was happening. The familiar logic of Silicon Valley—speed, scale, disruption—was suddenly in conversation with another worldview entirely: one grounded in lived realities, overlooked markets, and ways of solving problems that don't typically make it into boardrooms or moonshot roadmaps.

That moment was the enactment of the third space. A place where two different logics meet, unsettle each other, and make room for a new kind of possibility, one that couldn't have emerged from either side alone. And what emerged from that collision wasn't just a new strategy; it was, for me, a new kind of leadership potential. One that could hold different logics at once.

The 10X of the 10X.

At that moment, I realized that our current definition of 10X—the Silicon Valley version of exponential thinking—was too small. It captured technical ambition but not the full dimensionality of what moonshots now demand. What I witnessed in that room was the beginning of something bigger: the 10X of the 10X. A form of ambition that expands not just scale, but logic. Not just technology, but possibility. Not just performance, but perspective.

And here is the shift: the 10X'd 10X moonshot is an ecosystem moonshot—and it demands a different kind of leader. Not the founder who perfects a single product. Not the technologist who advances a single capability. Not the operator who optimizes a single business model. Ecosystem moonshots arise from the convergence of markets, communities, technologies, infrastructures, and incentives that rarely move together.

To lead at this level is to work across boundaries rather than within them—sensing across systems, building across differences, and projecting a direction of travel that harnesses the ecosystem around it.

The Paganini Stretch

So let's try it on for size and see what it opens in you and in me.

Before we move on, I want you to listen to something. Take a moment—even half a minute—and listen to Paganini's 24 Caprices, Op. 1: No. 1 in E major. Put on your headphones if you can. Don't think. Just let the sound come in.

Then return. (Nicolo Paganini, Itzhak Perlman)

Listening to Paganini's Caprice No. 1 is an exercise in expanded perception. The piece opens with a cascade of arpeggios that burst upward, forcing your ear to widen instantly. Multiple voices seem to emerge from a single violin—melody, counter-melody, threads of tension, moments of dissonance—each asking you to stretch your point of view just a little more. Paganini teaches you to hold more than one truth at once, to sit inside complexity without rushing to resolve it, and to let something unexpected and dimensional emerge from sound alone.

What Paganini does with sound, the next generation of moonshot leaders must do with perception, action, and projection. To lead from the third space is to work with complexity as material—to expand your field of awareness, hold different logics without collapsing them, and sense the deeper patterns emerging beneath the noise.

Just as Paganini creates the illusion of four violins playing at once, the ecosystem leader holds the distinct "voices" of the engineer, the community elder, the venture capitalist, and the street vendor. You aren't trying to make them play the same note; you are enabling them to play the same masterpiece.

Then, you move from that expanded awareness with clarity and intention, while projecting a direction of travel that reorganizes the possibilities around you. This isn't a leadership framework. It's an embodiment—a way of sensing, acting, and shaping the future that allows something larger to take form.

The following are the core capacities.

Self-Questioning Worldview

I want you to think about your worldview—how you believe the world works and what makes it work that way. Now think about all the experiences, the moments, the learnings, the struggles, and the data you've collected over your entire life that shaped that worldview.

Close your eyes for a moment and imagine looking at the world through that lens of yours—but only a small speck of the world is visible through it. Everything else is overcast. Hidden. Unseen.

Now imagine shifting your position. Turning your vantage point. Seeing the world from the north, the south, the east, the west—through the eyes of people whose realities are nothing like your own.

What begins to appear? What did you not know you didn't know? This is where ecosystem moonshots begin. They don't start with knowing; they start with not knowing.

The leaders who build them are the ones who can pause long enough to ask:

- How is my understanding of the world partial?
- What if my first instinct isn't the full story?
- What if the problem isn't where I think it is?

- What am I not seeing yet?
- Who or what will break open my worldview on needs, solutions, and opportunities?

Most people fear this kind of questioning because it threatens their expertise. But ecosystem leaders know something different: your certainty is often your smallest aperture.

The moment you unsettle your own assumptions, the system reveals what it's been hiding—the contradictions, the blind spots, the unexpected openings. This is the beginning of moonshot perception.

Anatomy of an Unlock: Harish Hande (SELCO Solar)

When Harish Hande began working in rural Karnataka in the 1990s, he carried a familiar engineering belief: that energy poverty was a technology problem. Solar panels were the solution, and people were poor because they lacked access to hardware.

Then he met a street vendor who shattered that worldview. She explained that she paid 15 rupees a day for kerosene—far more than a solar light would cost if she could pay in daily installments. The issue wasn't technology or demand. It was how her financial life worked.[1]

Harish later reflected: "Affordable doesn't mean low cost but appropriate financing."[2]

And in another interview: "We need to think why a person demands electricity.... It is always talked about from the supply side."[3]

In other words, his expert lens had blinded him to what was actually happening on the ground. People weren't "too poor" for solar—solar was too poorly designed for their realities.

He began listening differently: to farmers explaining income cycles, to women describing safety needs, to shop owners who lost customers without light, to communities who understood risk, credit, and trust far better than any bank model.

Their lived knowledge widened his aperture. His worldview shifted from "technology first" to ecosystem first. In 1995, he founded SELCO Solar—not as a hardware company, but as a decentralized energy ecosystem that combined:

- Locally appropriate solar solutions
- Financing aligned to daily and weekly cash flows
- Service networks rooted in community trust
- Livelihood models co-designed with users

Harish has said, "Our motivation is to help people out of poverty, not to be a leading technology company."[4]

Today SELCO has served more than half a million households and shaped India's national energy ecosystem. But the real moonshot wasn't solar panels. It was the moment Harish let other people's vantage points undo and expand his own. That is the **Self-Questioning Worldview** in action: the willingness to let your certainty become porous until a wider, more complex, and perhaps fuller reality comes into view.

Market Disobedience

Disobedience gets a bad rap. I like it. Because in the practice of moonshot building, disobedience isn't rebellion for its own sake; it's a disciplined refusal to inherit unquestioned market assumptions.

A third-space moonshot leader thrives in disobedience. Not to people, but to the logic that quietly governs what everyone assumes is possible, profitable, or plausible. These leaders have the courage to defy inherited market truths and create new economic openings the world hasn't valued yet. They are not investing in yesterday's market orthodoxy or in the "obvious" runways. They know the next great leap rarely starts where the market tells you it should. Instead, they seek out:

- Unseen and underharnessed ecosystems
- Silent signals that don't yet show up in reports
- Invisible markets hiding in plain sight
- Undervalued users and communities whose needs haven't been considered investable

To a third-space moonshot leader, untapped markets don't look risky—they look like discovery. They are the terrains most people avoid because they aren't "proven," or because they don't fit the expected sequencing of market entry. They don't match the pattern recognition investors use. They don't signal a quick win. And that is precisely why they matter.

Underleveraged markets are often dismissed because the risk–reward ratio appears predetermined. It's as if the market is saying: *We already know who the drivers are. We already know where value will come from.* But what if that "risk" is mispriced? What if the real risk is failing to look at these people, these needs, these ecosystems—and missing the multidimensional outcomes they make possible?

Market disobedience reframes the reward entirely: The reward isn't just financial upside. It's the creation of new categories, new economies, and new forms of impact that were impossible under the old rules. This is the heart of third-space moonshot leadership: refusing the inherited map so you can build the terrain no one has named yet.

Anatomy of an Unlock: Iceland's Geothermal Choice

Iceland offers one of the clearest demonstrations of market disobedience in the Global North. Long before renewable energy became an economic category, Iceland was quietly building a different future.

In the early 20th century, when global energy logic revolved around coal and later oil, the country began channeling its geothermal heat into public life. By 1930, Reykjavík had already launched a municipal district-heating system that piped geothermal water into homes, schools, and a hospital—a move almost no other nation considered technically credible or economically rational at the time.[5] And in 1969, decades before clean-energy investment curves took shape, Iceland activated its first geothermal power plant and expanded the idea of what a national energy system could be.[6]

When the oil crisis hit in the 1970s, the global energy order fractured. Most nations doubled down on oil imports or scrambled for short-term fixes. Iceland did something else: it accelerated its transition. The crisis became a forcing function. With oil prices spiking and vulnerability laid bare, geothermal energy shifted from an experiment to a national strategy.

Government programs, engineering research, and infrastructure investment surged. Instead of following the prevailing logic—*secure your oil, ride out the volatility*—Iceland committed even more deeply to a resource the world still treated as peripheral.

Today, geothermal energy provides the majority of Iceland's space heating and a substantial portion of its electricity, and the country exports geothermal expertise to nations across the world. Iceland didn't wait for a market signal; it built a market the world hadn't conceived yet. This is market disobedience at its clearest: refusing inherited value hierarchies, trusting the long arc of an overlooked ecosystem, and creating a future the market could only recognize in hindsight.[7]

Interdependence and Interconnection Instinct

Have you ever met someone who just needed other people or was fiercely interdependent? Without the tribe, they simply would not be themselves. And let's be honest: some of you who are not wired that way absolutely judge such a person. *Can't they take care of themselves? Can't they just stand on their own?*

But in the natural world, we see that what looks like "neediness" or mismatched pairing is often strategic. Some species seem completely at odds, yet out of necessity they become interdependent.

Take the unlikely alliance between tarantulas and tiny frogs. A tarantula is a predator; a small frog should be an easy dinner. Yet certain tarantulas allow these frogs to live in or near their burrows and even protect them from predators. In return, the frog patrols the nest and eats ants—tiny predators that would devour the tarantula's eggs. I first learned about this on my son's favorite show, *Creature Cases*, and assumed it was adorable fiction. Turns out it's true.

This is mutualism. And mutualism does not mean the benefits are equal or symmetrical. The interdependence is layered, uneven, and emergent. Each partner gets what it needs in a way that enlarges the possibility of the whole system.

A third-space moonshot leader operates from this deeper logic. They are constantly seeking how an ecosystem functions at the level of interdependence—where needs, opportunities, challenges, problems, solutions, and institutions converge in unexpected ways. You are sensing where a moonshot lives in interdependencies and unlocks a cascade of interdependencies embedded inside a problem or opportunity. It requires a felt understanding that no variable in the system is an island and that real moonshots are ecosystem events.

Anatomy of an Unlock: China's National High-Tech Industrial Parks In China's national high-tech industrial parks, innovation doesn't flow from isolated breakthroughs or the brilliance of any single actor; it emerges from how well the ecosystem learns to move together.

An empirical study measuring what the authors call *symbiotic efficiency* shows that regions where policy, firms, and technological infrastructures operate in genuine interdependence consistently outperform those where these relationships are fragmented. What mattered wasn't the presence of advanced technology or ambitious policies alone, but the quality of the reciprocal relationships among them—the degree to which each part of the system recognized itself as connected to and shaped by the others.

In the parks where those ties were strong, innovation accelerated; where they were weak, even abundant resources struggled to convert into meaningful progress. The study's conclusion is simple but profound: under rapid technological change, the most powerful driver of innovation is the strength of the ecosystem's symbiotic relationships. Interdependence isn't a soft value here. It is a measurable determinant of whether a region can achieve high-speed development or fall behind.[8]

Audacious Doubt

If you are a third-space moonshot leader, you must hold a dual truth: technology can profoundly expand human experience in ways we could not have imagined—and it can simultaneously cause harm, amplify inequity, and deepen existing problems. Holding both realities at once is not a contradiction; it is a discipline.

This is audacious doubt: the ability to believe in technology's potential while questioning, with equal intensity, what it unlocks, what it disrupts, and what it should never touch. It is doubt in service of the possibility of exacting interrogation—of where, when, and how technology becomes a genuine partner to ecosystem-level needs and solutions and when it must be deliberately left out.

Holding multiple truths is far more difficult than taking a singular stance. But it is the prerequisite for building more sophisticated technologies, tools, and interventions that are capable of operating at the scale and complexity of the world we now inhabit. Audacious doubt keeps us honest. It keeps us awake. And it expands what is possible with the right technology, in the best place, for the cascading outcomes.

Anatomy of an Unlock: The Robotics Bet I saw this play out while working with Intrinsic, the "Bet" on robotics and manufacturing. In the tech world, the concern is often binary: *will robots replace humans?* But Audacious Doubt pushed the question further. What if the bigger challenge isn't the hardware but the ownership of the change? Right now, automation often feels like it is coming *at* people rather than *for* them. At that moment, the "doubt" was about the logic of the rollout. If we build the most efficient system in the world but the humans within that ecosystem feel zero agency over the transformation, have we actually succeeded? Or have we just optimized a new kind of displacement?

Holding that doubt didn't slow the project down; it forced us to look at the cascading outcomes of how technology and human dignity must be integrated if the moonshot is to be sustainable.

The Science of Thinking Plurally Decision science reinforces this discipline. In *Multiple Perspective Dynamic Decision Making*, Leong argues that real-world decisions unfold under conditions of uncertainty where no single viewpoint is sufficient. Instead of collapsing complexity into one model, we must keep multiple perspectives visible and integrated over time.

Better decisions do not come from greater certainty, but from the capacity to hold and reason across competing truths. It asks us, as builders and leaders, to resist "premature closure"—to let technological possibility and technological risk sit side-by-side long enough to reveal what is genuinely wise. In a world of cascading consequences, this ability to think plurally is not hesitation. It is mastery.[9]

Multiscale Intelligence (Altitude + Ground)

The one question I heard endlessly at Alphabet was: *but can it scale?* I used to sit with that question and flip it on its head: what if scaling up requires us to first scale down? And when, exactly, should something scale? In the traditional sense, scale is about replication, reach, and whether a solution can be deployed across an entire organization or consumed by millions.

But in third-space moonshot leadership, we are not just talking about this traditional sense of scale. We are also talking about something else: the ability to move fluidly between vast ecosystems and perspectives and the nuanced realities on the ground and to traverse horizontally across contexts, regions, and timelines.

This orientation demands mastery across dimensions: strategic, operational, cultural, and experiential. It is the discipline of staying attuned to the macro forces shaping an ecosystem while simultaneously sensing the micro pressures, patterns, and constraints that determine whether a moonshot will take root and how it lives and succeeds.

True third-space leaders have the buoyancy to shift altitudes without losing coherence, to build solutions that work at the level of systems and the level of people. That dynamic movement—across scales, across contexts, across realities—is fundamental to inventing in a world defined by complexity, interdependence, and continuous change.

Anatomy of an Unlock: Indonesia's Tsunami Warning System Indonesia's tsunami early-warning system is one of the clearest examples of multiscale intelligence in practice.

At the highest altitude, the system integrates seismic sensors, deep-ocean buoys, satellite links, and hazard-modeling algorithms to read the Earth's movements in real time.

But none of that matters unless it translates into action at the ground level—families who know which routes to run, local leaders who understand which alleys flood first, and neighborhood networks that can mobilize in seconds.

The Indonesian system and global best practices stress that an effective early-warning system is a people-centered architecture: a synthesis of high-tech structural components and strong, well-trained community networks. The macro intelligence of ocean and atmospheric data becomes meaningful only when it is held together with the micro intelligence of local memory, lived experience, and communal coordination.[10]

This is multiscale leadership: the ability to build systems that see the world from orbit and respond with precision on a single street corner.

Aperture Expansion (Who Else/What Else Is Possible?)

How many people can you invite to your dinner table when you're trying to understand an issue? And equally, who is missing from this conversation? What else is possible that no one has named yet?

This is core to third-space moonshot leadership. It is the instinctive desire to widen the field with more voices, vantage points, and experiences, and, critically, a wider set of worlds entering the same space, rather than settling for the narrow band of perspectives that typically shape innovation. Bringing in these contexts is not an add-on or a courtesy. In third-space moonshots, it is the starting point.

A moonshot leader does not just ask the obvious questions. They ask:

- Have we looked at drone delivery through the lens of an astrophysicist and a community organizer?
- Have we explored commercial robotics from the vantage point of a biologist and a factory line worker?
- Have we examined machine learning through the eyes of an ethnographer and a mechanical engineer?

- What would this solution look like from Singapore? Lebanon? Alaska?

Aperture expansion is not simply about assembling expertise. It is about intentionally combining knowledge, perspectives, and realities that do not usually meet and allowing the friction, imagination, and surprise of that convergence to reveal something entirely new.

Beyond the Familiar Aperture expansion means saying: "Yes, we have five perspectives perfectly aligned with our moonshot area, and we also have five more from completely different domains. Now, together, can we ask: what else is possible?" If we believe the answer is X, is there also a Y or Z that becomes visible only when we stretch beyond the familiar? What happens when we see this problem from Manila's street level and through the eyes of a 10-year-old in New York? This practice pushes leaders to widen the aperture until possibilities long unseen, long unasked for, come into view. It is the moment when the forest and the trees shift—when we see them differently than we have for the last 100 years.

This process is relational and collective. It must happen in the first mile (opportunity reveals itself), the mid-mile (design solutions), and the last mile (deployment). This is not "more for more's sake." It is the recognition that when we stretch to seek the fuller range of human and planetary intelligence, we uncover deeper, more nuanced moonshot opportunities than would ever emerge from a closed or predetermined field.

Anatomy of an Unlock: The Ahmedabad Heat Action Plan (HAP) One example of aperture expansion in practice comes from Ahmedabad, India. Launched in 2013, the HAP was the first city-level preparedness system for extreme heat in South Asia.

The work began with climate scientists and epidemiologists modeling temperature spikes. But in a third-space move, they didn't stop at the "expert" data. They didn't just ask, "How do we scale our findings?" They asked, "Whose reality are we missing entirely?" They widened the aperture to include those whose lives are defined by the street:

- Street vendors and construction workers who occupy the heat all day
- Women in informal settlements who manage the household's water and health
- Grandmothers who held the "micro-data" of generations of survival

These communities named realities the models were blind to. They revealed that the "problem" wasn't just the temperature—it was the fact that shade disappears by noon in specific markets, that certain neighborhoods run out of water exactly when the heat peaks, and that children's behavior in the sun doesn't match the "safety protocols" written in an office.[11]

Once these worlds entered the design process, the entire definition of the solution shifted. It wasn't just a warning system; it was a redesign of urban life, water access, and community trust. They showed that scientific strength is incomplete without the field of voices. When the aperture widened, the solution changed shape.

Creative Dissatisfaction

Do you feel an impulse to build, especially to build what does not yet exist or what is still waiting to be discovered? Do you sense that if we created this one thing or reimagined a part of an ecosystem, people and societies could become something different, something other?

And equally, do you feel the quiet discomfort that comes from knowing that if we don't build it, we are silently accepting that life should remain as it is—when every part of you knows that life for people, ecosystems, and the planet could be different?

Creative dissatisfaction is the deep presence with an imaginative refusal to accept the present as the limit of what is possible for humans, societies, and the world.

That inner restlessness is not frustration or critique. It is not even hope. It is a life-giving refusal to accept the status quo when the world is calling for something other. It is the underlying pulse of a third-space moonshot leader: to open pathways toward what could exist. From this motivation, moonshot leaders are willing to take risks, to dwell at the edges where the human–technology relationship becomes an opening—an aperture through which this deeper motivation can be answered, so we can reach the center of something new.

Creative dissatisfaction is dissatisfaction in service of possibility, in service of human experience, and in service of futures that have not yet taken shape.

Anatomy of an Unlock: The Open-Source Insulin Movement I see this expression of creative dissatisfaction in the scientists and bio-hackers behind the Open Insulin Project. They asked a radical "third-space" question: what if the instructions and tools for manufacturing insulin belonged to the ecosystem? By doing so, they did not accept the status quo where a handful of companies control the global supply, relying on a massive, high-barrier manufacturing substrate that keeps production far away from the people who need it.

The Open Insulin approach is different. They didn't just want a cheaper product; they refused to accept that the "instructions" for a critical medicine should remain a proprietary secret. Their creative dissatisfaction was with the inaccessibility of the knowledge itself.[12]

They are currently in the thick of a working moonshot: developing an open-source protocol for "small-batch" insulin production. This requires navigating the dense interdependencies of protein purification, hardware design, and complex regulatory safety standards.[13]

Their work is a refusal to accept the existing logic of the market and the industry. This is creative dissatisfaction: the deep presence with an imaginative refusal to accept the limits of the current world, and the courage to build toward one we have not yet seen.

Can You Be a Third-Space Moonshot Leader?

Yes.

You see, the moonshot leader is the mirror to the moonshot opportunity. It's not just that the thing you build has to 10X; the "you" building it has to 10X as well. And that is not an arrival. It is an ongoing process of becoming so that the world can also become.

But you have to want this. You have to want to stand at the center of shaping the world. You've met people who move through a room with clarity, depth, and imagination, and you think, *I can feel their power*. What I want you to understand is that this is not a rare trait. It's a developed one. And it lives in you, too.

So as we close this chapter, you might be thinking: *how am I supposed to be all of this?* Here's the secret: you don't start as the person who can, and you don't do it alone. You grow into that person because the moonshot demands it. It stretches you, invites you, and provokes you into a bigger version of yourself. This is not a quest for perfection; this is growth with a sense of adventure.

A third-space moonshot leader learns to enjoy the altitude changes. You expand. You trip. You get up. You experiment. You surprise yourself. And in that ongoing becoming, you and the world become something other—something different. It is in that space, in that relationship, that moonshots emerge.

Moonshot Dojo

This is where we push and pull with the ideas in this chapter.

If you step into building something audacious, certain instincts—or trained ways of leading—will surface.

Good.

Now we spar.

Reflex: Of course my worldview is partial.

Translation: But I can already see more than most people.

Dojo move: Step into what you don't know.

Ask: If experience is an invaluable asset, what becomes possible when more of it shapes this moonshot?

Reflex: Let's keep this to real experts.

Translation: Experts + AI = we can model the world. Just.

Dojo move: Slip the expert-AI hold.

Ask: What might the world reveal that expertise and AI alone cannot see, experience, or know?

	Constrained Moves		Moonshot Moves
↛	Certainty	↚	Attunement
↛	Obedience	↚	Disobedience
↛	Stay in your lane	↚	Divergence

CHAPTER

5

From Tool to Partner

In the early 1990s, I found myself in the Northern Areas, a vast region folded between Pakistan, Iran, China, and India.

I still remember stepping off the plane, my feet touching the hot, energized ground. The smell of mulberry trees wafted in the air. Rivers roared as they cut through rock. Mountains became borders. Steam rose from tea and cooked meat at roadside vendors. A few men stood together, glancing down at pagers, silent in the mountains, waiting for signals. I was 19.

I had co-designed and was now implementing an educational development project for people living in remote villages with little access to advanced education. One of our guiding philosophies was simple: co-design and deliver with the people who would live with the system. In gratitude, the villagers took us on a picnic.

Families came. Babies wrapped against their chests, children darting ahead. Teachers. Elders. Goats. Baskets of vegetables. Small, makeshift barbecues carried up the mountain. We climbed narrow paths along the Hindu Kush mountain range, water running on one side, sheer drops of thousands of feet on the other, peaks rising well above 20,000 feet.

In the distance, white horses, sheep, and goats moved freely across the land. Apple and apricot trees filled the air with a tender sweetness. At one of the high points, I stopped, looked out, and gasped. I saw life differently. Not as isolated efforts or singular actors but as an interconnected ecosystem in motion. Every part in relation. Nothing extraneous or dominant or wasted. Each element—human, animal, terrain, culture—playing a role in sustaining the whole. It was one of the first times I understood how living systems work, not in theory but in the body.

Have you had moments like that? Where the vastness suddenly reorganized how you see things? Where the pieces revealed themselves as part of a whole? What did that feel like? What did it reveal?

Looking back, I now see this moment as one of the early building blocks that led me to what I now call third-space moonshot building. Not as a roadmap. But as an orientation. That experience gave me language for something I would later recognize again and again: the difference between designing from afar and designing in relationship to what already exists.

Third-space moonshots aren't something you follow step-by-step. They are something you learn to see, sense, and architect. And today, more than ever, we need people willing to stand at that threshold architecting human experience, life, and expansion across systems that touch hundreds of millions. We've moved from examining the Alphabet-era version of moonshot building to uncovering what a third-space moonshot is and what it requires to come into being.

So what do third-space moonshots do? They:

- Address real, unmet ecosystem needs with technology as a co-partner
- Trigger cascading outcomes that benefit hundreds of millions
- Reprice undervalued and under-leveraged markets by changing the conditions of value creation
- Unlock new, durable flows of economic (revenue and income) and societal value at key points in the ecosystem

Third-space moonshots don't chase scale. They reconfigure the conditions under which scaled impact with technology becomes possible. Even though in this framing of moonshots technology is not the center and instead a partner, it still matters, greatly. In fact, imagine your relationship with your partner or friend; in those relationships both you and they matter. It's together that you create outcomes, whatever they may be. I want you to consider technology as a partner of great significance. And like any partner there are times when you will, can, and should say no to it. But in this instance we are in the yes.

Technology Matters

Have you ever stopped to consider how often we underutilize or misutilize technology? The question can feel misplaced, because technology is no longer something we opt into. We already live inside it: on our phones, on our wrists, in our kitchens, and in the systems that determine loan approvals, healthcare diagnoses, and food pricing. AI, in particular, is no longer arriving as a discrete tool. It is becoming infrastructure, application, and interaction, shaping how life is organized, valued, and experienced. So the real question is not whether technology matters. It's where it actually changes what is possible.

Pause for a moment.

Where are you currently using technology simply because it's available and not because it meaningfully alters an outcome? And where might you be avoiding technology altogether, even though it could unlock something you cannot reach on your own?

Third-space moonshots ask technology to do something fundamentally different from some historical trends. Not unthinking automation. Not optimization everywhere.

Not substitution simply because it's feasible. They ask technology to operate at points of leverage where human effort alone cannot reach the required scale, speed, or complexity.

I was reminded of this recently at the dentist. The hygienist moved fluidly between a manual tool and an electric cleaner. She didn't default to the machine. She used it precisely where it reached places the manual instrument could not, where it accelerated completion, where it fundamentally changed what was possible in the same amount of time. The insight wasn't about the tool. It was about discernment. Knowing where technology alters an outcome and where it does not is the orientation of third-space moonshots.

Technology matters most when it enables outcomes that would otherwise be unreachable: synthesizing fragmented datasets, revealing patterns that would take years to surface, exposing disease pathways or systemic risks hidden inside complexity. In these moments, technology doesn't assist a system. It changes the conditions under which ecosystems can function.

Technology as Leverage

Here are two examples—one cultural, one commercial.

Think about Iron Man (yes, I'm a Marvel fan) not as superhero fantasy but as a system. What makes Tony Stark effective isn't raw strength or even the suit itself. It's how technology radically expands his ability to sense, decide, and act within complex environments: integrating information, compressing time, and widening the range of possible responses. The suit doesn't act on its own. And it doesn't replace him.

It changes the conditions under which action becomes possible. That distinction matters, because this is exactly what effective technology does in the real world without the spectacle.

Or consider navigation systems in modern cities. GPS doesn't make individual drivers better at driving. It reshapes traffic flows across entire urban ecosystems redistributing congestion, altering demand patterns, and changing how cities are experienced in real time. Although you still may end up in gridlock because there are human and structural dynamics GPS can't fix. In both cases, the value isn't automation for its own sake. The value is *where* the technology intervenes, opens up, or expands.

Try this lens for a moment. Think about a problem you care deeply about:

- What parts of the system are already working?
- Where does human effort start to break down?
- What signals exist that no one can see clearly?
- And where might technology—not as a solution, but as leverage—change the conditions of the whole?

Notice where the questions land you.

Recognizing Where Technology Will Change Ecosystems

More than six billion people are now connected to the Internet, with smartphone adoption accelerating fastest outside North America and Western Europe. Asia, Africa, and Latin America are not peripheral to the technified present or future; they are actively shaping it.

These regions are already defining new patterns of consumption, coordination, and social organization through technology. And yet, most technology systems are still designed around an outdated understanding of where economic agency sits. Capital, platforms, and AI tools continue to center users who represent a small share of future growth, while under-designing for populations that already anchor demand, adoption, and trust within their ecosystems. The issue is not just access. It is also orientation.

Certain populations are misread as only end users rather than ecosystem anchors—people whose daily decisions coordinate households, informal economies, and social infrastructure. Because their influence is diffuse rather than institutional, their economic value is often underestimated or ignored. This is where some third-space moonshots begin.

Look around your own community. Who are the anchors? The people making healthcare decisions for families. The ones moving goods from place to place. The quiet educational gatekeepers. The informal communication networks that determine which services are trusted and used.

When technology meets these populations as a co-partner, supporting how they already coordinate money, care, information, and labor, the effects are not incremental. They are multiplicative. Small interventions at these key points ripple outward, stabilizing households, increasing participation across local economies, accelerating adoption, and reshaping market behavior at scale. This is not an argument about identity or development. It is an argument about leverage.

Women, in all their plurality, are one visible example of this broader pattern. Globally, they influence or control a majority of consumer spending and are poised to receive trillions of dollars in accumulated and inherited wealth in the coming decades. At the same time, especially across the Global South, many remain structurally under-supported by financial, technological, and AI systems, not because they lack economic potential but because that agency does not conform to traditional models of market power.

Third-space moonshots do not aim to reinforce existing systems or optimize outcomes those systems were designed to produce. They identify places where outcomes are structurally constrained, where ecosystems underperform not due to lack of talent or demand but because the surrounding conditions cannot support what could be emerging and who could benefit.

In these contexts, technology is not a solution in isolation. Used as a co-partner alongside human, economic, cultural, and institutional forces, it enables new outcomes to replace constrained ones; these are outcomes that propagate through households, supply chains, and adoption networks already operating at scale. As these outcomes take hold, systems reorganize around them as a consequence, not a goal.

In doing so, hundreds of millions of people are impacted. Technology becomes an extension of human coordination and decision-making, right-sizing economic, and societal expansion so that progress and well-being can scale together rather than compete.

This is the difference between marginal improvement within existing constraints and outcome-level shifts that reorganize participation, value creation, and what is possible at scale.

Looking Where Others Are Not

It's compelling when searching for a moonshot solution to look where success already is, where people are making progress, and where there is momentum, capital, and validation and to find the angle no one has tapped into yet.

But what if you looked where others are not? At your garbage can. Your dirty laundry.

The vegetable fiber you throw away after making a smoothie. Learning to see and work at the neglected or undervalued points of an ecosystem is a core practice of third-space moonshot building.

In many ways, AI has made building possible for more people than ever before. And yet, we still tend to focus on the main ingredients, the obvious inputs, rather than what happens to what is deemed waste. Part of this is cultural. Part of it is economic. And part of it is technological: incentives still reward building bigger, better, newer, rather than re-seeing what already exists.

But what if the most valuable opportunities lie precisely in what is overlooked, undervalued, or thrown out?

Pull out your phone. Just stare at it for a moment. What does it hold? What is being used? And what is quietly accumulating in the background?

In 2025, the average smartphone user spends roughly four to five hours a day on their device and consumes on the order of tens of gigabytes of mobile data each month, a figure expected to continue rising through the decade. There are now more than five billion smartphone users globally and even more devices in circulation. And yet, only a limited portion of the data our phones generate is actively used to help us live better or make more informed decisions.[1]

Much of it sits elsewhere:

- App telemetry and crash logs
- Location pings from background services
- Search and browsing history
- Notifications, drafts, and unsent messages
- Unused photos, screenshots, and traces of intent

A small share of this data is used to keep networks functioning, optimize performance, or improve a handful of visible features. The rest is effectively *personal dark data*—collected and stored, consuming energy and resources, but rarely translated into insight, foresight, or improved outcomes.[2] Now multiply that pattern by billions of people. Globally, mobile networks already carry well over 100 exabytes of data each month. Across organizations and sectors, industry research consistently suggests that a large portion of collected data, often more than half, is never actively analyzed or used. It is gathered, stored, paid for, and then largely ignored. This is not just a consumer phenomenon.

Governments face similar challenges, with many datasets under-used, inaccessible, or poorly integrated into decision-making. In health research, long-standing concerns about incomplete reporting mean that substantial portions of research outputs fail to fully inform practice or policy, reducing the value of massive global investments.

Even AI infrastructure itself produces excess. Data centers generate enormous amounts of heat simply by doing what they do—processing, computing, running nonstop. Most of that heat goes nowhere. It's vented, dissipated, and treated as an inconvenience rather than a resource.

But when you start to look more closely, something interesting happens. In the right context, that excess heat can be captured and put to work, warming buildings, supporting local energy systems, and offsetting demand that would otherwise rely on fossil fuels. What was once a byproduct becomes a contributor.[3] What we treat as exhaust often holds the signal. The real question isn't just what AI can do for us. It's how much latent capacity already exists and how quickly it can be transformed into shared capability and resources.

A growing layer of dark data is now being generated by AI systems themselves—logs, intermediate representations, feedback signals, and synthetic outputs produced, stored, and often used only for narrow operational purposes. Each interaction leaves behind traces of how people actually work, learn, coordinate, struggle, and adapt. Most of this intelligence is used narrowly: to optimize engagement, tune models, or hit internal KPIs.[4]

But what if it were used differently?

Within systems, could this data help shape safer workflows, better education tools, earlier health triage, or more responsive public services? Across systems, machine-generated data from factories, farms, hospitals, logistics networks, and cities, could data, if responsibly governed and pooled, strengthen climate modeling, infrastructure planning, and disease surveillance?

This is what third-space moonshots ask us to see. Not only what we need to add but what is already here, mispriced, underused, or ignored. Technology, used with discernment, does not extract more from systems. It changes what they are able to produce. What once looked like waste becomes leverage and opportunity. What sat in silos became shared infrastructure. Constrained outcomes give way to entirely new ones.

Any Technology for a Third-Space Moonshot?

One of the things that most fascinated me about the Alphabet moonshots was the depth of commitment to technological breakthroughs. There was a seriousness to the work, a willingness to pursue problems so complex they bordered on science fiction. The goal wasn't incremental improvement; it was to push into territory that didn't yet exist.

As a lifelong lover of science fiction, I found that deeply compelling. There was something exhilarating about being around teams willing to sit with uncertainty for years, sometimes close to a decade, just to get the technology to a place where it might work at all. In one case, an engineer told me they had spent seven years refining a core system before even beginning to think about commercialization. That kind of runway is unique. And it also reveals something important. Traditional moonshots are built around long arcs of technological maturation. The breakthrough *is* the work. The technology itself is the mountain that must be climbed before anything else can happen.

The third space is different. Here, technology is not the destination. It is the enabler. What matters is not whether a technology is entirely new but whether it can act as a catalyst at the right moment, with the right people for outcomes that benefit—whether it can unlock a 10X or 100X shift by amplifying something that could be inspired to, already wants to, or needs to move.

In the third space, technology can come from many places:

- A breakthrough finally ready for application
- A tool built for one purpose that quietly outgrew it
- A discarded or underused system waiting for a new context
- An existing capability combined in a way no one had previously considered

What unites these is not novelty. It's possibility.

AI, in particular, can dramatically shorten parts of the development cycle. It can accelerate pattern recognition, simulation, testing, and iteration. Processes that once took years can now unfold in months. But acceleration is not the same as readiness and strategic application and adoption. Even early research on AI use, such as Anthropic's Economic Index, shows that the technology's impact comes less from the tool itself than from how and which people integrate it into work, systems, and everyday decisions.

AI does not collapse the entire arc of development, deployment, and impact. It cannot replace trust, judgment, governance, or the social conditions required for technology to land, adopt, and create beneficial outcomes. It does not eliminate the work of integration or the failures that become invaluable lessons. So while AI can compress parts of the journey, it cannot compress the whole of it.

That's why the central question in third space is not: what technology should we invent for X problem? It becomes: what is now possible and what kind of technological partner would allow that possibility to take form? That shift from invention to possibility, from breakthrough to leverage, is where we find and pursue third-space moonshots.

The Conditions That Allow Technology to Work for Us

I often find myself in conversation between two poles: those who build technology in pursuit of scale and those who focus on what is local, lived, and immediate. In the world of third-space moonshots, the work happens in the space between. What ultimately determines whether technology creates value or distortion is the degree of proximity it maintains to the realities it is meant to serve. And whether reality serves it. Whether they can be in a symbiotic and evolving relationship.

There is a prevailing belief that effective technology requires distance—objectivity, scale, abstraction. We're taught to step back, to think in markets and metrics, and to ask what will sell, what will scale, who the user is in the aggregate. These questions aren't wrong. But when solely centered, they pull us away from something essential.

What often gets lost is intimacy, closeness to the people, contexts, and conditions a technology must engage and serve. The kind that comes from being near enough to notice and perhaps even be in the experience of friction, contradiction, and nuance. Where someone will tell you why you will get this wrong in their neighborhood or city. Not as a UX exercise but as a revelation. Where you will learn that the infrastructure won't support the tech, not because it's not good but because decision-makers won't align. It is near enough to see where real needs diverge from assumed ones. It is near enough to recognize when a solution solves the wrong problem elegantly.

Third-space moonshots insist on that proximity. It treats closeness not as bias or simply as subjective data but as a signal, as a compass. It understands that many of the most consequential insights don't emerge from distance or scale but from sustained contact with the lived realities of people, of systems, of contexts. Without that closeness, we optimize for abstraction. With it, we design for relevance. This is the tension at the heart of moonshot innovation: the pull between scale and specificity, between distance and intimacy.

So what does this mean in practice?

Most technologies are designed to do something very well, to solve a defined problem within clear parameters. They excel at repetition, precision, and optimization. Some can even operate in complex environments: like drones navigating uneven terrain, processing massive volumes of data, or robots simulating physical and biological systems with extraordinary accuracy.

But what technology continues to struggle with is the messiness of lived reality. When people enter and build systems, things stop behaving cleanly. Variables multiply. Incentives blur. What works in theory bends under real-world conditions. And it's here, in this space between precision and unpredictability that many technological approaches falter or produce undesirable outcomes.

Consider how often we try to predict behavior using data. We want to know: will this person buy that razor, choose that drink, or subscribe to that streaming channel? And once they do, how do we ensure they keep doing it?

It sounds simple.

But what happens when the razor you normally use gives you a cut and you decide to wax your legs instead? Or when the show you've been watching suddenly brings up grief about your mother passing and you stop watching anything that reminds you of it? Or when a doctor makes a decision about your care based on clinical data but before taking the prescribed medication you talk to your local aunties and learn that this doctor has a reputation for ignoring side effects?

In moments like these, the tidy logic of prediction begins to unravel. And it's here, in the space between precision and unpredictability, that many technological approaches begin to falter or produce outcomes people may not have chosen.

This is why proximity matters. Being close enough to see how people actually live, decide, and adapt reveals frictions no model can predict. And it's here where we can ask, what role can technology play? Where might it help us navigate, address something with us? And it's in navigating this dynamic, between what machines can optimize and what human systems actually require, that third-space moonshots grow.

Consider something as ordinary as a school bus. In most cities, school buses run for a few hours in the morning and afternoon and then sit idle for the rest of the day. They are maintained, fueled, insured, and distributed across neighborhoods with remarkable precision. Yet for most of their lifespan, they are treated as single-purpose assets. Seen through a third-space moonshot lens, this changes.

Those buses represent latent infrastructure—mobile, trusted, and already embedded in communities. In some contexts, they could support energy resilience, act as mobile charging hubs, or serve as connective tissue during emergencies. Not because someone invented a new technology but because someone noticed what was already there and asked a different question.

The insight isn't technological. It's relational. It comes from proximity from seeing what a system already does, who it already touches, and where unused capacity quietly exists. This isn't a proposal so much as an illustration—a way of noticing how much latent capacity already exists in systems we take for granted.

Leaping with the Leaps

If a school bus is latent physical infrastructure, then artificial intelligence is our latent cognitive infrastructure. It is the most significant "co-partner" we have ever built. But it is leaping with a velocity that can feel overwhelming—moving from chatbots to autonomous agents to physical robotics in a matter of months. In the face of that speed, it is easy to default to a mixture of frantic adoption or paralyzed distance.

In the third space, we don't just use the leap; we learn to leap with the leap.

This requires a fundamental shift in our orientation. Most traditional innovation is a "static launch": you build a platform, you aim, and you fire. But when the technology itself is accelerating while you are building on it, you cannot afford a static launch. You need kinetics. Think about the slingshot metaphor: in this era, the tension is being re-cocked while the projectile is still in the air. This is the "infinite spring." To navigate this environment, we must embrace a new ruleset for discernment.

I advised an educational technology company that has spent decades building and delivering digital learning content. Like many organizations at this moment, they are searching for their next leap. Their instinct has been to respond to the moment by producing AI-related content—domain-specific training and development designed to keep pace with changing expectations.

There is nothing inherently wrong with that approach. But it is still using the leap to maintain relevance.

Leaping with the leap would start somewhere else entirely. It would ask a different question: not who needs to learn about AI or what new content to produce but how humans will come to know things, build skills, gain knowledge, and develop capability in a world where AI is woven into everyday life. From that vantage point, the real question becomes what role this organization should play in shaping the educational ecosystems that will build societal capacity in the decades ahead.

That means thinking beyond content toward nimble educational infrastructure, talent systems, platforms that adapt and long-term capability building—toward the nodes across sectors and regions that will help define how knowledge, skills, and opportunity move through society.

This is what it means to leap with the leap.

Velocity Matching: Architecting for Outcomes

Have you ever met someone you just can't keep up with? No matter how fast you walk, they are always 20 paces ahead. Every time you try to close the gap, you just end up feeling more behind. This is the "AI trap" we are currently in—trying to sprint alongside a technology that is fundamentally designed to accelerate beyond us.

Human velocity cannot match AI's, and it shouldn't try. Instead, our task is to get very comfortable with its changing trajectory. The mistake is designing for what AI *is* or what it *does* today. Because what it does will change by next Tuesday. If you build your moonshot around a specific feature, you are building on shifting ground. Instead, a third-space moonshot is designed for the outcomes that AI helps make possible.

While the "what" of the technology fluctuates, the "what" of our desired outcome—introducing a platform that cascades to new economics or an infrastructure that makes data an invaluable asset—remains our north star. Our job is not to keep pace with the software updates; it is to keep pace with the evolving potential to produce the results we want. We aren't chasing the tool; we are harnessing its momentum to reach a destination we couldn't reach on our own.

The New Standard

To believe that AI is the final frontier of human change is to mistake a star for the galaxy. What the current moment reveals is not simply the power of large language models but the velocity and inevitability of the forces now reshaping our systems.

We have entered an era where the "server" of human knowledge is no longer a static library but a self-accelerating engine. If we fixate solely on AI, we are essentially building a compass that only points north while the very poles of our reality—biological, material, and social—are already shifting beneath us.

The shift we are witnessing through AI is a kind of kaleidoscope for what is coming in synthetic biology, quantum materials, and decentralized energy. We are moving from a world where we *use* technology to one where technology increasingly becomes the operating system of our physical existence.

Just as search engines quietly re-architected our view of ourselves—how we store, consume, and access knowledge—the coming wave of genomic editing and molecular manufacturing will re-architect our agency in other ways: what humans can design, repair, grow, and create.

When we begin to recognize that these technological and human shifts are not isolated events but overlapping transformations, the question in front of us changes. The task is no longer simply to invent new tools. It is to reimagine the ecosystems we will live within. And this will continue to be true in the decades ahead as there will be evolutions, breakthroughs, and advancements we have not named yet or imagined.

For third-space moonshots, this opens up a view of possibility that is moving how we architect the conditions of the futures we will inhabit together and that will themselves morph.

The Nonlinear

AI doesn't yet "think" in the human sense; it calculates probabilities across trillions of connections that no human brain can hold at once. When it suggests something "nonlinear," it is simply exploring a mathematical space that we don't have the "eyes" to see.

AI can find a connection between a "mulberry tree" and a "supply chain algorithm" purely through data patterns that are invisible to the naked eye. It provides a provocation from an angle that doesn't exist in the standard human range of view. This is the ultimate "Yes, And": the AI offers the randomized spark, and we offer the discernment to see if it lights the way to a new market or a more durable flow of value.

Human Intelligence as the Stabilizer

As the tech leaps at 20X, 100X, and 1000X, the human provides directional stability. AI provides the propulsion; we provide the intent, the placement, and the lived experience. The embodied intimacy of being human. Without our proximity to the "Hindu Kush" of the ecosystem, AI cannot and perhaps should not be more than a force.

We are living in a moment where the "dark data" of our lives, and the latent heat of our systems are waiting to be reclaimed. The leap is already happening. The question is: are we building moonshots catching the wind it creates? And will you choose technology as a partner?

Moonshot Dojo

In a dojo, your partner is not simply someone you spar against. A partner is a force you learn to move with. Through that exchange, you begin to experience what neither of you could do alone.

Our relationship with technology is often far less graceful. Some days we love it. When it meets our needs, we're all in. When it starts making decisions for us—or behaving in ways we don't understand—we're ready to throw it out of the house. So we are about to explore this relationship in a way that makes even me a little uncomfortable. I'll admit that. It requires stepping out of the abstraction that often shapes how we relate to technology and engaging that relationship more directly.

Technology deserves that kind of attention. Its implications are too great for a casual connection. If it is going to be a partner in your third-space moonshot, you must move beyond the user manual and into a practiced relationship check-in.

Vulnerability Check

What parts of my—or our—thinking, actions, evolution, or experience should you change—and where do I hesitate to let you shape what comes next?

Agency Check

In this partnership, how will people still have agency?

Will we feel empowered—or something else?

How will we navigate the complexity that comes with building something together?

Outcome Check

If I use you in this way—if I place you here—what will unfold for me, for others, and for the systems we live within?

Will you make us different?

Better?

Will you take something away—something we need to release, or something we should keep or protect?

Will you benefit some while harming others?

And will the world you help create be one we actually want to live in?

System Check

If you become part of __________ ecosystem, what relationships shift?

Who or what gains new capacity?

What new value begins to circulate?

These questions—and others you will develop—are a way to pause and spar with the partnership you are forming with technology before the momentum of building carries you forward.

And in a third-space moonshot, this is not something you do once. It is a practice you return to again and again.

CHAPTER

6

Markets as a Moonshot

We were on a call with investors for a moonshot I co-founded in low-altitude air space intelligence, including drone detection. We had laid out the problem, the need, and the approach. The conversation was steady, interested, but measured. Then I shared the total addressable market (TAM), serviceable available market (SAM), and serviceable obtainable market (SOM).

I watched the shift happen in real time. Pupils widened. Postures changed. Chests subtly expanded. The energy in the room recalibrated. The numbers were big—multimillions, even billions—but they weren't inflated. They were grounded in reliable market data. What surprised people wasn't just the size of the opportunity. It was how we were getting there. When we described our beachhead—an off-the-beaten-path entry point—people leaned in.

"Huh. That's not how anyone else is approaching this."

That moment crystallized something important for me. When we talk about markets in the traditional sense, we're usually talking about scale: big numbers, existing demand, accepted categories. Markets are framed as pools of money already waiting to be accessed—where customers are known, sales cycles are predictable, and logic is inherited.

But in third-space moonshots, the market itself is part of the moonshot. Take Waymo. Mobility and transportation were pre-existing markets. But a world of cars that drive themselves did not yet exist. It had to be constructed. Not just technologically, but socially, regulatorily, culturally, and economically. The market had to be built both within and beyond what already existed.

In traditional moonshot thinking, the market is often addressed after the technology is ready to land. In third-space moonshots, we work with a different trifecta from the start: market, technology, and impact (benefits), designed together, in relationship with one another.

When we were thinking about our drone-detection solution, it wasn't just about identifying and tracking drones. It was about recognizing how airspace itself becomes a ripple opportunity. Airspace is not a single-use market; it is an ecosystem. One that, when approached thoughtfully, enables far more than what you start with.

This is where third-space moonshots fundamentally diverge. The market is not a transactional input–output equation. It is a dynamic ecosystem—one that asks different questions:

- Where could money flow that currently doesn't?
- Where could access expand?

- Where might opportunity compound in places most people overlook?

We've already talked about undervalued and underleveraged markets. This is where that idea deepens. This shift isn't just about economics; it's about power. Who has it now. Who could have it next. And how markets, when designed as ecosystems, can make people and solutions not just viable but indispensable.

To see this clearly, we need to loosen our grip on some inherited market logics. The first sounds like this: *if X number of people have problem A, and we have solution B, then the market is what people are willing to pay to solve that problem.*

The second shows up in a familiar pattern: The spark: "AI could optimize irrigation schedules." The build: 18 months and $2 million later, a sophisticated system emerges. The reality: farmers don't have Wi-Fi in their fields—or the budget for a subscription.

The failure here isn't technological per se. And it isn't that the market doesn't exist. It's that the market was misidentified. The solution was built as if the farmer were the buyer, the field the point of leverage, and the subscription the value exchange. But the real opportunity could live elsewhere in the ecosystem—where water risk is priced, compliance is enforced, or yields are underwritten. This is what happens when we apply otherwise sound market logic to the wrong place.

We cannot all be Waymo. We don't all have a decade, unlimited capital, and regulatory patience while the world catches up to our technology. By braiding market, technology, and impact from day one, we aren't building for a future that might arrive. We're building solutions that create their own markets as they move.

The Shift

Third-space moonshot markets don't reject traditional market dynamics, but they recognize them as more limiting than what's possible. Traditional market dynamics operate like a revolving door. People walk in and out. Value is exchanged on the other side. Something is gained. Something is lost.

In third-space moonshots, the market is not simply utilitarian or transactional. We're looking for a different set of signals.

Traditional market lens: Who is the buyer? What is the product? How big is demand? Third-space market lens: Where does value stall? What relationship governs that stall? What switch—if flipped—changes flow for everyone?

Third-space moonshot markets emerge not just from unmet demand but from misaligned relationships that prevent value from flowing. They are not defined by people or institutions alone. They are shaped by what exists between them. Ecosystems don't fail only because the wrong actors are present. They fail because the spaces between actors are misaligned, underdesigned, or never intentionally designed at all.

Traditional market analysis focuses on nodes—buyers, sellers, firms, agencies. Third-space builders look for something else entirely: relational infrastructure. The unseen places where opportunity either compounds or collapses.

These show up as gaps in coordination between public and private actors. As trust deficits between communities and institutions. As regulatory gray zones that stall innovation without formally banning it. As data that exists, but cannot legally, ethically, or operationally flow. None of these looks like a market. Yet each one acts as a source.

I don't want to suggest that flipping these switches is easy. It isn't a light touch; it is a structural heave.

In a traditional market, friction is treated as an obstacle to be bypassed or optimized away. In the third space, friction is where the market can emerge. If flipping the switch were a simple flick of the wrist, someone would have done it already. These switches remain "Off" because they are hot—wrapped in political risk, long histories of mistrust, or labyrinths of legal liability that no one wants to touch.

They are often rusted shut by decades of "that's not how we do things," held in place by gatekeepers who benefit from the stall or by agencies so fragmented that no one has the authority to move the lever. Flip the switch, and value moves. Ignore it, and even the most elegant technology stalls.

To flip a switch in the third space, you have to do more than build technology. You have to build the legal, social, and ethical scaffolding that makes it safe for the switch to move at all. You aren't just discovering a market; you are absorbing risk that the market was too afraid to carry. This is the unbillable work—drafting protocols, rebuilding trust, and creating liability shields that allow value to circulate.

That's why third-space moonshot markets rarely announce themselves through customer surveys or demand curves alone. They reveal themselves through persistent friction—friction that refuses to resolve despite talent, capital, and intent being present on all sides. Where friction persists, value is often trapped. And where value is trapped, markets may be waiting to be born.

A Drone Example: When Flow Restores the Market

Consider the emerging need for drone detection and airspace sensing in urban environments. Traditional market analysis asks: Who is the buyer? Governments? Stadiums?

Corporations? What will it cost? What is the return? How many units will they purchase?

These aren't bad questions. But they miss something essential. The constraint is not demand. It is not technology. It is not even cost. It is a set of inactive flow switches.

In most cities, responsibility for low-altitude airspace is either nonexistent or fragmented across agencies. Liability is unclear. Sensor data doesn't flow easily between private property owners and public safety entities. Communities don't trust surveillance systems. Regulators haven't defined acceptable use but haven't prohibited it either.

Nothing is wrong enough to force action. Nothing is clear enough to enable investment. So value stalls. Now imagine activating just one flow switch: a trusted, privacy-preserving mechanism that allows anonymized airspace risk data to move between private sites and public authorities without creating surveillance or liability exposure.

Suddenly:

- Property owners gain situational awareness.
- Cities gain early warning without new infrastructure.
- Insurers can price risk.
- Security providers unlock new services.
- Standards bodies have something real to regulate.

A market appears—not because a product was sold but because flow was restored. The opportunity was never the drone. It was a stalled opportunity in the airspace ecosystem.

This is why emergent markets are often a signal that a moonshot is nearby.

They require us to imagine not just a product but a shift in how people live, move, exchange, and coordinate. When a true third-space market begins to form, it quietly asks a different question: *if this existed, what would change about how the world works?*

Not in theory, but in daily life. How people insure risk. How responsibility is shared. How decisions are made. How value moves between actors who were never designed to interact.

Emergent markets don't appear where the world already knows what to do. They appear where the world hasn't yet agreed on the rules because those rules haven't been written. That uncertainty is often mistaken for "no market." In reality it can be a signal that something new is trying to come into form.

Moonshots live here because they demand more than execution. They demand projection—the ability to see ahead of current behavior into a future that doesn't yet have language, infrastructure, or business models, but where momentum is accumulating.

How I See These Markets

There's a practice I return to again and again when I'm trying to understand where a real market might live. I sit down and visualize the landscape—not the pitch-deck version but the lived one.

I imagine a need and a potential technological partner. Then I scan the terrain. I look for the obvious nodes: companies, agencies, customers, regulators. Those are easy. Everyone sees those.

Then I slow down.

I pay attention to the in-between spaces. Where energy stalls. Where responsibility diffuses. Where trust matters. Where value wants to move but can't.

I imagine placing the technology here. What shifts? Who becomes newly powerful? What friction softens? What cascades?

Then I move it there. What breaks? What accelerates? What unexpected alignment forms? What divergence brings new opportunities?

I'm not asking where the product fits. I'm asking where flow changes. And it's usually in that moment—when a dynamic reconfigures rather than a transaction completes—that a market begins to make itself known.

This isn't a mental model; it's a practice.

I did this once in conversation with the CEO of one of Alphabet's Other Bets companies. We were talking about expansion into emerging markets—places where the need was undeniable but entry points were complex. I suggested starting in under-served or deregulated regions, where impact would be immediate but possible only through partnership with a global entity like the United Nations.

He stopped me. "We're not doing humanitarian work," he said. "We need paying customers. Business partners."

So we slowed the conversation down. We looked at the United Nations (UN) not as a moral actor but as an operational one. A purchaser of goods and services. A coordinator of logistics. A supplier embedded across regions with active cash flows in the hundreds of millions. An institution already sitting inside the very markets this technology was designed to serve.

More importantly, we looked at what the UN unlocked: Regulatory clearance. Local trust. Speed of adoption. Regional legitimacy.

Not charity. Flow. And yet, he couldn't see it. In his mental model, "commercial" and "noncommercial" were mutually exclusive categories. Markets were places you sold into, not systems you activated. The idea that a nonprofit institution could function as a market switch simply didn't compute.

The conversation ended there. The opportunity didn't. The market wasn't missing. It was misrecognized. That's why third-space moonshot markets so often appear where others insist there is "no business case." The case exists. The flow does not, yet.

Once you learn to see ecosystems this way, your job as a builder shifts. You stop asking, "Who will buy this?" You start asking, "What has to be unlocked for this to work and create value across points, people, and outcomes?" That question doesn't just change where you build. It also changes what you recognize as a market at all. And once you see markets this way, you start finding them in places no one taught you to look.

Moonshot the Money!

At some point, every serious moonshot runs into the same question: where does the money come from?

This question is framed too narrowly. We treat it as a funding problem, a pitch problem, a capitalization problem. For third-space moonshots, money is not the starting constraint. It is the visible signal of something deeper: how wealth exists and money exists, how it flows, and who has access to it.

To understand why some moonshots are possible and others never leave the ground, we have to distinguish between money and wealth.

Money is liquidity. It is cash, runway, budget, burn. It is time-bound and finite. It runs out. Anyone who has built outside large institutions knows this intimately. Every line item matters. Every month carries risk. Scarcity is not theoretical; it is lived.

Wealth is different.

Wealth is accumulated capacity. It is infrastructure, assets, optionality, trust, and economic and cultural currency and continuity. Wealth compounds. It absorbs risk. It allows failure without collapse. It creates the conditions in which experimentation is survivable over time.

When I worked inside Alphabet, I saw this distinction clearly. The company didn't simply have money. It had decades of wealth—products that functioned as durable assets, business lines that continuously generated surplus, institutional trust that reduced friction, and systems designed to carry risk forward rather than extinguish it.

The Other Bets existed inside that wealth.

Even when individual bets felt pressure, scrutiny, or existential urgency, they were operating within an ecosystem already rich in foundations. The irony was structural. There was an acknowledgment of abundance, yet leadership often felt pushed up against scarcity—responsibility to boards and employees. That pressure was real. But the underlying conditions were not the same as those faced by builders operating outside institutional wealth.

I felt this myself working inside the Other Bets system—a kind of cushion that was always there. Even when a company ended, the cushion remained. Living inside that kind of abundance permanently changes how risk and possibility are perceived.

For those building world-changing solutions without that foundation, scarcity is not psychological. It is structural. It shapes what can be imagined, what can be attempted, and how long a team can stay with uncertainty before being forced into premature closure.

I have lived this reality too as a moonshot founder. I have been on the funding circuit—sitting across from venture capital firms, private equity, family offices, personal family capital—translating a vision to fund. I have felt the quiet pressure of questions about timelines, exits, patents, and comparables. I have watched imagination narrow when capital is asked to behave defensively.

These are structurally different realities.

And yet third-space moonshots demand something audacious: ecosystem-level change must be matched with ecosystem-level wealth and money. That doesn't mean what you might assume. I am not arguing that moonshots require being Google, a government, or an institution with access to millions or billions of dollars. In fact, that assumption has narrowed the creative genius moonshots demand and constrained the possibilities they are meant to unlock

And this is where most funding advice fails.

The dominant narrative assumes a narrow path: raise capital, exchange equity, scale fast, exit. That model has produced extraordinary companies. But it is not the only pathway through which wealth and money can be activated, nor is it always the right one for ecosystem-level moonshots.

Third-space moonshots require a different relationship to money—not as a reward for success but as infrastructure for emergence.

Here is the shift:

> Money does not fund moonshots.
>
> Moonshots reorganize money.
>
> Capital is rarely absent. It is stalled. Under accessed. Unknown.

Money accumulates around clarity, legitimacy, and flow. When it does not move, it is usually not because the idea lacks merit but because the system surrounding it cannot yet absorb the risk, responsibility, or consequence of letting it move.

This is why so many founders exhaust themselves chasing capital, when the real work is reorganizing the conditions that would allow capital to circulate.

If you are building a moonshot capable of reshaping an ecosystem, you cannot approach money as something scarce, adversarial, or morally suspect. Nor can you approach it with entitlement. You must stand in the reality that capital exists—and that it is seeking coherence, stewardship, and durable value creation at scale.

Money flows toward structures that can hold it responsibly.

That does not mean capital is neutral or benevolent. It means it responds to architectures that reduce systemic risk while expanding possibility. A third-space moonshot leader must embody that readiness—psychologically, ethically, and structurally.

This also means expanding where wealth itself is generated.

Markets, when designed as ecosystems, do more than enable transactions. They create new forms of assetization: shared infrastructure, data commons, risk-pooling mechanisms, trust networks, and coordinated value flows that did not previously exist.

These are not abstractions. They are the foundations of wealth.

When wealth is created at the ecosystem level, money does not need to be extracted from a single buyer or captured through a single product. It can circulate across actors, compounding value across time and across lives.

This is what it means to moonshot the money itself.

Not by inventing new financial instruments or mimicking venture-capital logic but by changing the conditions under which value becomes visible, investable, and durable.

By designing systems where more people participate in wealth creation—not as charity, not as inclusion theater, but as a structural outcome of how the ecosystem is organized.

I saw this most clearly while working on our drone-detection venture.

At one point, a state official told us that funding was limited due to federal budget cuts. That statement could have ended the conversation. Instead, it clarified the system.

If money was blocked at the federal level, the issue wasn't scarcity. It was location.

So we asked a different question: "Where is money already paying a price—without visibility, coordination, or return?"

Municipalities absorbing infrastructure liability.

Property owners carrying security risk.

Insurers pricing uncertainty without data.

Communities living inside airspace complexity without representation.

None of these actors was a "buyer" in the traditional sense. But all of them were already paying—in inefficiency, exposure, and unpriced risk.

The opportunity wasn't to sell a product. It was to unlock a stalled flow of value. This is the kind of strategic gymnastics third-space moonshots require to exist at scale. Not to find someone who can fund the whole thing but to identify where money is frozen—and what would allow it to move.

The same logic applies to wealth more broadly.

There is another structural shift happening in parallel—one most innovation ecosystems still fail to recognize.

Today, women control roughly one-third of global investable assets—an estimated $50–60 trillion worldwide—and that share is growing rapidly.[1] In the United States alone, women control more than $18 trillion, projected to exceed $30 trillion within this decade as intergenerational wealth transfers accelerate.[2]

This is not marginal capital. It is one of the largest reallocations of wealth in modern history.

And yet most of this capital has historically sat outside the core networks that fund technological innovation—not because of lack of capacity but because the structures of venture and private equity were not built in relationship with how these asset holders assess risk, time horizon, or impact.

The same is true for many people who accumulate wealth outside technology itself: founders who exited non-tech businesses, family enterprises, real-estate holders, professionals, diaspora communities, inheritors. Collectively, these groups represent tens of trillions of dollars that rarely flows into moonshot-level ventures.

What this reveals is not a shortage of capital. It is a misalignment of pathways. A disconnect between those who could help shape the next generation of ecosystem moonshots and those positioned only to receive the technologies that transform their lives—sometimes even absorbing their harms.

To build third-space moonshots—moonshots capable of unlocking entire ecosystems of value, opportunity, and flow—know this: a much wider group of people already hold the power to shape the next generation of moonshots. Ecosystem-level moonshots require ecosystem-level participation in how wealth is created.

The money is already here. The work is to build the pathways that allow it to move.

This is about meeting ambition at the scale money actually operates.

Moonshotting the Money Experiment

Up to this point, I've called you into a vision: the world needs a collective of people to build third-space moonshots. Naturally, your brain is firing off "But how?" flares.

Who pays for this? Doesn't this cost millions—or billions—of dollars? Historically, moonshots have required governments, Big Tech, or the kind of elite pedigree where a billion-dollar outcome seems obvious. Think of the Apollo missions, the Internet, or global computing infrastructure. Big money. Big resources.

But we are not here to replicate. We are here to moonshot the moonshot.

Let's run a thought exercise together. Not because I'm giving you a blueprint, but because I want to see what happens when we turn our creative prowess "on."

Move: The Discovery

Imagine a need, an opportunity, or a systemic "glitch" you've discovered alongside a collective of people from wildly different domains. You aren't building a startup with a few co-founders; you are discovering a moonshot within an ecosystem of people, contexts, and experiences.

This is something that:

- Touches multiple systems (e.g., food, energy, and labor)
- Crosses national boundaries
- Requires many actors to coordinate
- Produces cascading benefits if solved

Maybe it's building data infrastructure for shared airspace to coordinate threat response. Maybe it's addressing how human labor could be cross-border amid massive shifts in education and talent markets. In the traditional model, addressing this would require years of engineering, proprietary systems, and tens of millions of dollars before the first person ever experienced the solution.

Move: The Variable Shift (the $100,000 Trigger)

Now, imagine a different scenario. Imagine the technological layer of your solution costs $50,000. Or $100,000.

Not $10 million. Not $1 billion.

This number is not a prediction; it is an observation of a collapsing cost curve. In the traditional model, value was trapped in the software and the talent. Today, what once took an engineer years to build can be assembled by nonengineers in weeks.

The Bridge and the Car

To understand how this experiment works, we have to look at two layers of the technology landscape.

- **Foundational technologies (the bridge):** Frontier systems—large AI models, advanced chips, satellite networks. They cost billions and take decades to build. You are not building the bridge.
- **Applications (the car):** Once the bridge exists, you build the vehicle that travels across it. You assemble tools, models, and platforms that already exist. Today, that might mean accessing a billion-dollar intelligence system through an API call costing only $0.15.

In some cases, you may build outside the bridge—using open systems, local models, or lightweight infrastructure. But even then, you are drawing on layers of capability that have already been built.

The pattern holds.

The bridge expands the range of journeys possible. But the car is what makes the journey real. The bridge is imperfect—often built without the world's full scope of genius. Still, we strategically use it as a trampoline.

Move: The Capital Pivot (Liberating the $9.9 Million)

If the technological layer now costs only $100,000, we are left with a provocative question: what do we do with the other $9.9 million?

In a traditional moonshot, that money is "captured" by a single entity to pay for proprietary code. In a third-space moonshot, we liberate that capital. We take the millions that would have been burned in the furnace of R&D, and we reinvest it into the nodes of the ecosystem. Now imagine your solution across an ecosystem.

It reaches farms, small businesses, city systems, schools, and community networks. But technology entering a system is not the transformation. Solutions must still be adopted, integrated, adapted, and refined through real-world use. As people engage with the solution, the system begins to respond. New uses emerge. Adjustments are made. Connections form between actors who were previously isolated.

The solution evolves through its relationship with the ecosystem. And as this unfolds, something else happens.

The surface area of the moonshot expands. More people and more places reap the benefits of the original moonshot and also gain the ability to invent solutions to the needs around them—and to the needs they share with others across borders and boundaries. When the technological layer becomes easier to assemble, we can afford to build solutions where the benefits are not captured by a single platform. Instead, value appears across many points in the ecosystem.

We move from building technology to regenerative orchestration. This investment flows into:

- **Domain expertise:** Paying the farmers, nurses, or transit workers to design the workflows they actually use
- **Institutional capacity:** Strengthening the schools, clinics, or cooperatives so they can host the solution
- **Trust architecture:** Building the governance and protocols that allow disparate actors to coordinate without friction

Move: Distributed Skin in the Game

If we aren't waiting for a $10 million check from a single source, where does the initial investment come from?

In a third-space moonshot, investment could be distributed. Because the cost of the technological trigger is low, the barrier to entry drops. This allows different actors to "invest" their specific resources into their piece of the ecosystem.

How the Ecosystem Invests

- **The practitioners (workers/users):** They invest their time and domain expertise. They participate in the design phase to ensure the solution actually solves their pain points. Their "cost" is the labor of adaptation.
- **The institutions (schools/local government):** They invest in infrastructure and legitimacy. They provide the physical spaces, the existing networks, and the regulatory "green zones" where the experiment can live.
- **The collective:** They invest the Initial trigger capital and integration talent. They aren't building a fortress; they are building the "car" and teaching others how to drive it.

In the traditional model, you pay with your agency. You get a "free" tool, but a platform provider owns your data and captures the value. In the third space, you pay with your shared genius. You take on an initial cost—time, focus, or local capital—but in exchange, you own a piece of the upside.

Move: The Shifting Calculus of Value

When the cost of the "car" is tiny compared to the value of the ecosystem, technology becomes scaffolding. It is no longer where value accumulates; it is what activates value flows.

As a solution is adopted and refined through real-world use, value begins to circulate in ways that are not captured by one actor:

Value Category	The Downstream/Upstream Benefits
Income and Livelihoods	Profits move to different people and nodes in the ecosystem.
New Markets	Orchestration creates "niches" for local entrepreneurs to build new services.
Knowledge and Data	The ecosystem "gets smarter" together; data is a shared asset that yields profits for many through access fees and use subscriptions.
Environmental Restoration	Benefits (soil health, clean air) accrue to the land where the actors actually live.
Resilience to Shocks	Distributed systems don't break as easily, so resilience is built into communities to mobilize with change. Cost are reduced for key nodes and actors.
Capacity	New capabilities begin to emerge and grow stronger and more capable of activating opportunities and solving the next problem.

Move: The Calculus of Circulation

If the technology is no longer the moonshot, then the calculus of the moonshot must change.

In a third-space moonshot, success is no longer measured only by profit captured or technology deployed. It is measured by how much value continues to open and circulate through the ecosystem.

We can express this shift through an equation like this:

$$\text{Ecosystem Unlock} = \frac{\text{Value Created} \times \text{Value Circulating}}{\text{Coordination Friction.}}$$

Where:

- **Value created:** The output generated by the ecosystem—solutions, services, income, profits, knowledge, and new capacities.

- **Value circulating:** Regenerative yield—the multiplier effect as value spreads and compounds across the ecosystem over time.
- **Coordination friction:** The dynamic of aligning actors across the ecosystem: trust-building, governance, integration, adoption, and the distributed effort required to enable the opportunities and change.

The Traditional Model

Spend $10 million building a proprietary solution.

If the company succeeds, it may generate significant financial returns for investors and founders.

The success metric is clear: how much value did the company capture?

The Third-Space Model

Spend $100,000 to assemble the technological trigger.

Now invest the rest in activating an ecosystem where time, resources, infrastructure, profits, and capital begin flowing through multiple actors.

In the first cycle, let's say the ecosystem generates:

- $6 million in new local income
- $2 million in city savings
- $2 million in new local services

Total value circulating: $10 million.

But unlike the traditional model, this value is not the end of the story. As the ecosystem continues to evolve—new actors joining, new services emerging, new knowledge accumulating—the value compounds.

Now keep moonshotting this thought experiment.

It is imperfect. It has gaps. So turn it on its head. Push on those gaps. See if they lead you somewhere entirely different—or keep building in this direction.

The Third-Space Insight

Today, our ability to generate intelligence is expanding. But the opportunity is no longer simply to think more powerfully—it is to link and engage intelligence across ecosystems.

Technology will continue to advance. Breakthroughs will continue to emerge.

But the next frontier of moonshots will not be defined by the code alone.

The breakthrough isn't the code.

The breakthrough is the ecosystem.

The real question is no longer whether we can build the tool. It is whether we are willing to invest in the orchestration that allows an ecosystem to become—unlocking cascading value that expands what becomes possible.

Third-space orchestration is a wager: that a thriving ecosystem is more valuable than a proprietary tool. That cascading outcomes—income, livelihoods, knowledge, resilience—are not features of the system but signals of its expansion and evolution.

This is not simply the financing of a solution. It is the underwriting of a new collective capacity. Technology may spark the possibility. The ecosystem it activates is the moonshot.

Moonshot Dojo

Most people treat money as a constraint.

In a third-space moonshot, money is also something you redesign.

So go ahead—moonshot the money and the market surrounding it.

CHAPTER 7

The Capacities Moonshots Now Require

At least a few weekends a month, my parents would lay down mattresses in the living room of our two-bedroom apartment. Our extended family would come over and stay the weekend. All 10 of us would be huddled together—aunts, uncles, cousins—eating ice cream sandwiches. Bollywood music played in the background. The smell of chai and Indian snacks filled the air.

And then, like clockwork, it would happen. One uncle would clear his throat and claim the attention of the room. Then, in a magnanimous voice, he'd say, "Bacha (children), remember this—one stick can be easily broken, but many sticks together are often unbreakable."

We'd glance at each other—roll our eyes, this again?—and then launch into a flurry of questions. What if one stick breaks in the bundle by mistake? What if one of the sticks rots? What if something stronger comes along and breaks the whole bundle?

My uncle would look at us, faintly irritated by the tussle. We hated and loved that story. It was the same every time—the same tone, the same enthusiasm. A lesson that replayed like a song you couldn't get out of your head. And somewhere inside us, we knew there was something more trying to reveal itself. So we'd smile and go back to playing.

Later in life, I began to understand the mathematics behind what my uncle was saying. At the heart of human progress lies a profound paradox: the power of the group is never merely the sum of its individual parts. It is a multiplier. I was reminded of this when I came across the work of the fourteenth-century philosopher Ibn Khaldun, who wrote that "what is obtained by the cooperation of a group of human beings satisfies the need of a number many times greater than themselves."[1]

This insight suggests that intelligence is not a private resource to be hoarded, but a shared infrastructure—one that generates a surplus of value. When we transition from the individual mind to the collective intellect, we move away from isolated problem-solving and into a field of emergent wisdom. What I didn't realize at the time was that my uncle and all of us kids were engaged in what Aristotle called entelechy: the idea that what emerges through interaction is greater than what exists in isolation. What arises out of the intellectual tussle is shared intelligence.

The Art of Intellectual Aikido

This "tussle" my cousins and I engaged in—the questioning of the bundle, the poking at the rot—was actually the beginning of a practice I call Intellectual Aikido.

In traditional innovation, we try to solve problems through raw force: more data, more compute, more capital. Aikido, however, is the art of using an opponent's momentum to create a new direction. In a moonshot, the "friction" is the energy of the system itself. Instead of trying to ignore the rot in the bundle or bash through the disagreement, we use collective intelligence to absorb that tension and turn it into kinetic energy. We don't just tolerate the tussle; we use it to pivot.

Collective Intelligence in the Third Space

In the third-space moonshot, collective intelligence is not just valuable. It is indispensable and perhaps even primary. Third-space moonshots are not powered by individual brilliance, but by collective intelligence that has been intentionally activated. Not crowdsourcing or consensus or "wisdom of crowds" clichés. But situated, relational, multiscale intelligence that no single actor can hold. And where there is a knowing that we will find and build a moonshot through intentionally activated, collective intelligence across global contexts, brilliance, and life.

The Supercharge

Have you ever been at a concert or anywhere a large group of people gathers and felt the energy in the room turn electric? What did that feel like to you?

Everyone is vibing. Each person moves in their own way. Their voice and tone are their own. Their rhythm is distinct. And yet, together, something palpable emerges, something no individual could create alone.

I have to admit: every time I imagine a global collective building a moonshot, I feel like I can fly. There is a fire in my belly—excited and alive. The realization that we could address a real need with expertise, experience, and insight drawn from many places and many lives feels like a supercharge.

In the traditional moonshot, an individual or a very small group sometimes produces extraordinary breakthroughs. But what if we could exponentially expand not just who builds moonshots, but how they emerge? What becomes possible when intelligence doesn't stop at the individual but begins to proliferate through a collective grappling with problems together—mapping ecosystems, seeing from multiple angles and altitudes at once?

The supercharge doesn't come from everyone agreeing. It comes from the dance between divergence and convergence. Perspectives rubbing against one another fast enough to generate heat and light. Divergence without fracture. Convergence without flattening. Collective flow, with tension, is kinetic. It's the difference between a pool of still water and a high-pressure jet. The goal of the collective is not to arrive. It is to activate.

Alignment is the moment when technology and humanity collide with enough force to send a wave through an entire ecosystem transforming a single breakthrough into many new possibilities.

The true moonshot is not a terminal point on a map. It is a catalytic event. When we align collective friction, blending human intuition and knowledge with technological scale and systemic capacity, we don't just aim at a target. We create a pulse. We drop a stone into the center of the third space and watch ripples of opportunity move outward, unlocking doors we didn't even know existed.

The Next Leap in Moonshots

The world is saturated with intelligence. Experience. Expertise. Knowledge. History. Experimentation. Lived application. Intelligence that emerges through relationships, geographies, cultures, and systems far beyond any single institution or discipline.

Yet much of what we call "moonshot thinking" has quietly narrowed this intelligence. Traditional moonshots often privilege a limited set of perspectives, credentials, and forms of expertise, operating as if only certain kinds of intelligence are capable of producing breakthroughs. That assumption may be one of the very things holding us back.

Given the complexity and interdependence of the challenges we now face, what lies ahead is not a refinement of the old model but a more radical opportunity. Climate instability, public health crises, fractured information ecosystems, uneven technological benefit, and geopolitical complexity do not yield to brilliance in isolation. Nor is the present explained, or the future defined, only by the problems we must solve. It is also shaped by the opportunities we create for collective flourishing—where human ingenuity, and perhaps even the "soul of the world," expresses itself through discovery, creativity, enjoyment, in the presence of what life is and in the expansion of what life can be.

These are ecosystem-level challenges and opportunities. They require intelligence that can operate across scales, contexts, and lived realities at the same time.

The next leap in moonshots is not about assembling smart teams to build ever-smarter tools or technologies. It is about building collective capacity—multi-layered, multi-perspective intelligence grappling together to reach solutions and untapped possibility no one could arrive at alone.

This does not always require technology, nor should it. There are moments when restraint is the more intelligent choice

For decades, we have invested in incubated spaces of intellect and expertise—labs, accelerators, research groups, and elite teams designed to concentrate capital, credentials, and technical capability. These spaces have produced breakthroughs. But they are optimized for depth within bounded domains, not for intelligence at the scale of whole ecosystems.

What we need now are connective spaces—third spaces—designed to optimize ecosystem-level intelligence for ecosystem-level needs and opportunities. Spaces capable of holding capital, infrastructure, and institutional authority alongside local knowledge, lived experience, cultural context, and operational reality not as inputs to be integrated later or that we add but as intelligence that how we understand what is needed, are understood and how solutions and opportunities are discovered and designed from the start.

In these spaces, the work is not a consensus or coordination. It is the harder task of increasing the system's overall capacity to see, interpret, and respond to complexity. The goal is not agreement, but fitness: intelligence that can operate across scales, adapt under pressure, and generate solutions no single vantage point could produce alone.

The future of moonshots will not be determined by who controls the most advanced tools. It will be shaped by who knows how to mobilize intelligence that already exists, distributed, plural, and deeply embedded in the world it is meant to serve.

The next leap in moonshots changes the unit of innovation from individual brilliance to collective capacity. Collective intelligence is not a feeling. It is a practice, one that requires structure, discipline, and the willingness to stay with complexity longer than comfort allows.

The Global Unsiloing

A friend of mine, a virologist, once shared something that has stayed with me, partly because of how much frustration sat underneath it. He described how, in virology, the dominant instinct is to isolate. Isolate the pathogen. Isolate the organ. Isolate the cell.

This approach has driven enormous progress. But he was increasingly uneasy with what it leaves out. What if the source of disease is not located in a single place? What if it emerges from the interaction between systems—biological, environmental, and social moving together?

Before COVID-19, asking that question was often intellectually tolerated but structurally unsupported. Research lived behind paywalls. Data access was gated. Disciplines advanced in parallel rather than in conversation. The system rewarded depth within silos, not visibility across them.

Then the virus arrived.

COVID-19 did not behave like a problem that could be isolated. It moved through bodies, cities, supply chains, travel networks, media ecosystems, and political systems at once. It did not slow down for peer-review cycles. It did not wait for institutional coordination. It exploited every gap between how knowledge was produced and how the world actually works. And under that pressure, the rules broke.

Research that would normally remain locked behind journals was released. Data crossed borders that had previously constrained collaboration. Scientists worked across fields not because collaboration had suddenly become virtuous but because containment was impossible otherwise.

The virus was the catalyst. It forced a temporary global unsiloing, revealing something that had long been true but rarely acted upon: some problems do not live in parts. They live in interactions.

COVID-19 was not a failure of science. It was a stress test of scientific architecture. And it exposed the limits of a world still organized around isolated intelligence, when the challenge itself was fundamentally systemic.

What emerged was not a finished model but a glimpse of a different one—a recognition that when problems operate at ecosystem scale, intelligence must move differently. Faster. Wider. Across domains that were never meant to remain separate.

This is what I mean by *global unsiloing*—not as an aspiration but as a structural response to problems and needs that cannot be contained.

AI and the Cost of Isolated Intelligence

Recently, I was on a panel about the opportunity of AI across multiple sectors. Afterward, a medical doctor came up to me, visibly frustrated, and asked a question that captured the problem perfectly.

"Why are there so many AI companies knocking at my door to sell me tools," she said, "when they don't understand that the problem I'm solving is not a single-factor authentication code?" Her work, she explained, does not live in one system. It sits in the space between patients, data, research, clinical judgment, institutional constraints, and care delivery.

Yet the solutions being offered to her were designed as if her reality could be reduced to a narrow technical function. "They're trying to fit a solution into a system they don't truly understand," she said. "Or maybe don't even know they need to understand."

What made her frustration more acute was that she was not resistant to AI. She knew she would need to adopt it in some form. Not doing so could put her practice at risk. It could even impact patient outcomes. But adoption under pressure is not the same as adoption with understanding.

Like COVID-19, AI is exposing the limits of intelligence developed in isolation. The failure does not sit in the algorithm alone. It emerges at the interfaces between technology and medicine, optimization and care, speed and responsibility.

The AI Paradox

This is the same structural fault line we saw during the pandemic. We have built tools that are the ultimate "isolators." AI can optimize a single function with terrifying and exhilarating speed, but today it lacks the capacity for metamorphic intelligence. It doesn't yet "think" in the third-space sense; it predicts based on the past and what it might generate. It cannot "grapple" or "wrestle" with a reality that isn't already in its training data. When we mistake its *prediction* and insight for human *perception*, we create a gap where the system fails.

Global unsiloing is beginning to surface here as well, not because institutions are ready but because the cost of not doing so is becoming impossible to ignore. We may need to start thinking of global unsiloing not as an organizational preference but as a requirement for continued human and planetary progress. This is not about building a bigger brain but about building a better-connected nervous system.

For too long, we have treated collective intelligence as a social virtue—something admirable, optional, even a little kumbaya. In reality, it is becoming a computational requirement. The moonshots of this decade, and the ones to come, are simply too large, too fast, and too interconnected to be held by isolated intelligence, no matter how well-resourced.

The pandemic functioned as an alpha test for planetary-scale intelligence, revealing both what becomes possible when knowledge flows freely and how brittle our systems remain when it does not. What comes next is not speculative. It is the work of making that intelligence operational.

To do that, we need to be more precise about what we mean by intelligence in the first place.

In technology, we have overwhelmingly privileged technical intelligence—what can be computed, optimized, and scaled. That intelligence matters. But it is not sufficient for the world we are now building in.

In a third-space moonshot, intelligence does not live in abstraction alone. It emerges through an embodied relationship with the world as it is actually lived—through what is tangible and intangible, visible and invisible, measurable and felt.

This requires other forms of experience—ways of seeing, listening, sensing, and interpreting context that allow systems to be understood as they behave, not just as they are modeled.

When these forms of intelligence are held together—technical, situational, relational, and embodied—we get something different. Not more intelligence, but intelligence that operates across states at once.

That is the intelligence required for the next leap in moonshots. And it does not arrive as a single capability. It shows up as three distinct forms of intelligence that must work together if ecosystem-level moonshots are to take shape.

Translational Intelligence

I was on a high-level panel at the United Nations, in conversation with leaders from across the globe. I had brought several of my technology colleagues with me. Almost immediately, I noticed something familiar. People were speaking earnestly to one another but in fundamentally different languages—shaped by different contexts, assumptions, and incentive structures. The techies were saying one thing, the diplomats something else. Meaning was lost between them.

On the surface, that shouldn't be remarkable. Of course, people from different domains speak differently about the same thing. But I've come to see moments like this as far more consequential than they appear.

What breaks in these rooms is not just intelligence. It is movement.

Again and again, I see insights stall at boundaries—between technology and policy, medicine and data, infrastructure and lived reality. Not because people lack expertise but because the capacity to carry meaning, constraint, and responsibility across domains has not been built. That capacity is what I call translational intelligence.

Translational intelligence is not communication skill or stakeholder alignment. It is not simplification, dilution, or making ideas more "palatable." It is the ability to move intelligence across domains without stripping it of context, agency, or consequence. It works through differing incentives, logics, and constraints, allowing multiple vantage points to shape both the understanding of a need and the design and adoption of an ecosystem moonshot solution.

This is what was missing in the conversation with the doctor who approached me after that AI panel. What she needed was not another tool, but translation: a way for clinical reality to shape technical design, and for technical possibility to re-enter care delivery without erasing judgment, responsibility, or trust.

Without translational intelligence, adoption becomes coercion. Solutions land outside of relationship. Intelligence on all sides goes underutilized or distorted. What looks like resistance is often a failure of translation.

I have sat in front of rooms filled with hundreds of people and shared this insight, and I almost always see the same response: recognition. Not surprise, recognition. A collective nod to something people already know, because they live it every day.

We intuitively understand the value of translational intelligence. What we have not done is recognize it for what it is: a core intelligence we must intentionally build capacity for, if we expect complex systems to change without eroding trust or agency—and to build solutions that are genuinely responsive to everyday life.

In third-space moonshots, translational intelligence allows intelligence to move across sectors, scales, and systems without breaking apart. And if it breaks, it's an opening for something else. It is what turns collective intelligence from an aspiration into an operational foundation.

I want to really press this: what if we built a moonshot that was truly a cross sector, cross scale, cross experience moonshot from the start. What would we create? What would emerge? Yes, yes, all your buts are coming up, but I need you to move with those buts, like an Aikido practice. You are not pushing against, you are moving with the flows of resistance, misalignment, difference into other forces. You have to embrace the friction as the opening.

Metamorphic Intelligence

In the world of Alphabet moonshots, the one thing I learned early on was to become comfortable with death and failure—for ideas, projects, and businesses to die not as an end state but as part of the process.

In innovation more broadly, failure is often framed as an opportunity for something else: iteration, adjustment, learning. Third-space moonshots require something more than that.

This is not simply about being adaptive. It is about intelligence that emerges from sustained engagement with problems and opportunities that are genuinely hard and deeply complex. Intelligence that learns how to morph in response to what it encounters—to step back or forward, to move sideways, to circle until a new path reveals itself. To let ideas, inputs, and solutions die and be reborn so something else can grow.

This kind of intelligence does not resolve quickly. It grapples. It wrestles. It refuses premature closure. It lives in a productive restlessness—one that continues evolving until new needs, opportunities, and solutions come into view for a moonshot that is waiting to take form in an ecosystem.

I recognize how difficult this is. This way of working does not lend itself to efficient, streamlined approaches to technology or clean narratives of progress. It resists certainty.

And certainty is what most people want—clear definitions, linear paths from problem to solution, confidence before commitment. I understand that impulse. But complexity does not yield to certainty. It yields to capacity.

What I am arguing is that the next leap in moonshots requires us to genuinely embody ongoing change, adaptation, and detachment—in our thinking, in our approaches, in what we build, and in what we believe we can change.

Enough to stay with uncertainty without freezing. Enough to let forms fall away when they no longer hold. Enough to move with a problem as it reveals itself, rather than forcing it into shapes we already recognize.

Metamorphic intelligence is not comfortable. And it is not for everyone. But it is what allows ecosystem-level moonshots to emerge when old forms are insufficient and something fundamentally new must take their place.

The story of how the Grand Ethiopian Renaissance Dam came into existence offers a compelling example of metamorphic intelligence in practice. Ethiopia knew from the outset that traditional channels of international infrastructure financing were unlikely to support a project of this scale, given the complex geopolitical interests surrounding the Nile Basin. Rather than abandoning the ambition, the country grappled with that constraint and transformed the problem itself.

The project ultimately became a nationally financed effort—supported by Ethiopian banks, civil servants, merchants, and diaspora communities. One of the largest hydroelectric dams in the world emerged not through conventional development pathways but through a reconfiguration of ownership, financing, and national participation.

In doing so, Ethiopia did more than solve an infrastructure challenge. It reshaped the logic of how such a truly ecosystem level need could come into being—demonstrating how systems that appear stalled can be outcreated when new forms of collective capacity take shape.

This is what metamorphic intelligence looks like at scale: not simply adapting within the existing system but reshaping the conditions under which something new becomes possible.

The political and economic negotiations surrounding the dam continue to evolve, as they inevitably do in projects of this scale. But that ongoing entanglement is not a failure of the process. It is part of it. Metamorphic intelligence does not eliminate complexity; it learns to move with it.

Situational Intelligence

If you've grown up in a city like Toronto or New York, you learn to observe, engage with, and even become part of the streets you walk. You learn the directions of the roads. The landmarks. Which streets curve and which lead to dead ends. Where traffic flows and where it reliably clogs. You register the constant noises, what moves above you, how culture shifts from block to block. You come to understand the infrastructure—what services exist where, which streets are supported, and which quietly absorb neglect. This is what we often call situational awareness.

But situational awareness is not just about what you notice as you move through a place. It is not a detached scan of your surroundings. At its best, it is the recognition that markets, cities, communities, and geographies are alive—constantly signaling what is working, what is strained, and what is breaking down.

Situational intelligence takes this one step further.

It is the ability to read those signals clearly: where systems are failing or only partially functioning, where people are thriving despite constraints, where gaps are structural rather than superficial, and where blind spots persist because no one has stayed close enough to see them. It is learning to observe not just static conditions, but the dynamics and behaviors of people, infrastructure, and institutions over time.

Situational intelligence is the capacity to let context teach you—revealing where a real need exists, where intervention is possible, and how a moonshot solution might emerge from the conditions themselves, rather than being imposed upon them.

I saw this firsthand in the urban airspace intelligence moonshot I co-founded. As we spoke with communities, government officials, law enforcement, and commercial operators, a realization kept surfacing. People would instinctively look up and ask a simple question: why have I never really thought about what is above me?

I remember sitting with a law enforcement chief—an incredibly thoughtful and experienced leader—who said aloud what many people quietly sensed but had never articulated: how have I not seen the sky as a space of information?

As those conversations unfolded, the airspace most of us experience only as background—where we hear birds, planes, or the occasional drone—began to feel different. It became raining signals: concerns about safety, questions of authority, economic incentives, community expectations, lingering data, and pathways for coordination. The sky itself started to reveal layers of need and opportunity.

At that moment, airspace no longer felt empty. It felt like living data—valuable geography that had been present all along, but that few people had learned to read.

Data plays a critical role here but not in the way it is often treated in technology systems. Data does not speak for itself. It reflects what a system, and those building within it, is able to see, measure, and value at a given moment in time.

Without situational intelligence, data becomes a proxy for reality rather than a window into it. Metrics harden around what is easiest to capture, while blind spots accumulate around what is lived, contextual, or politically inconvenient. Patterns appear stable even as underlying conditions shift.

Situational intelligence determines which data matters, which data misleads, and when new forms of data—qualitative, local, relational, or real-time—must be brought into the system before intervention is responsible.

Without situational intelligence, moonshots scale too early, land in the wrong places, or solve for abstractions instead of lived realities. With it, interventions align and opportunities can cascade.

Intelligence for the Next Leap in Moonshots

What this chapter has been arguing is not easy to accept. The constraints facing moonshots today are no longer primarily technical. They are not a lack of talent, tools, or data.

They are constraints of intelligence itself: how it moves, how it adapts, and how it responds to the realities it encounters. When intelligence is isolated, unseen, and denied even, it stalls and shrinks. It can fracture, calcify, and even misfire.

The next leap in moonshots will not come from adding more intelligence into the system. It will come from changing how we engage with intelligence across domains, scales, and lived realities.

Translational intelligence allows intelligence to travel. Metamorphic intelligence allows solutions to creatively evolve. Situational intelligence allows intervention to emerge from and land in the messiness of real life.

Together, they shift the unit of innovation from individual brilliance to collective capacity. This is not a call for better collaboration or broader participation. It is a recognition that the problems now before us demand intelligence that can operate as a living system, able to move, adapt, and respond without collapsing under its own complexity.

The question, then, is no longer whether we can imagine bigger moonshots. It is whether we are willing to be shaped by the forms of intelligence required to shape them into the next leap.

Take a beat and listen to the "Equator Song" by Nyokabi Karuiki.

Moonshot Dojo

Bring one need, opportunity, or problem on to the mat.
Name it.
Now run three moves.
Push.
Pull.
Redirect.

Push—Translational Intelligence

Move the idea into another domain.

What's getting translated that wouldn't otherwise?

Pull—Situational Intelligence

Let the environment act on the idea.

What is revealing itself now?

Redirect—Metamorphic Intelligence

Now introduce complexity.

Turn the idea sideways.

Where does it want to move now?

Ask: Now, what else is possible?

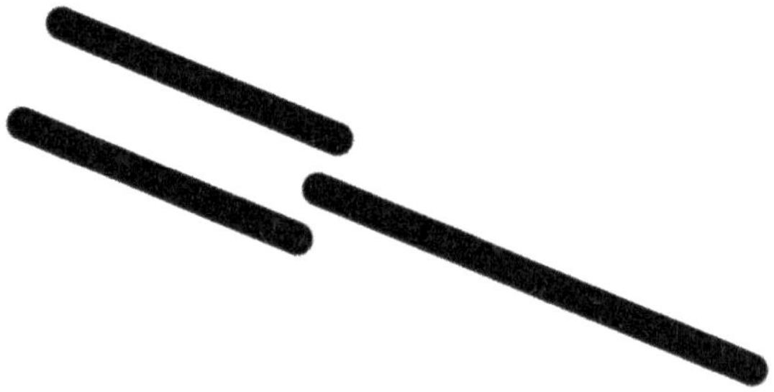

CHAPTER 8

Choice

At this point in the book, you are in the stew of moonshot making—in the mix of 10X'd ambition, systemic friction, and the pressure of building in the third space. You've encountered new ways of seeing markets, intelligence, scale, and ecosystem-level possibility. The nonlinearity of building a third-space moonshot may leave you feeling energized, unsettled, resistant, or eager to move quickly—to apply, to build, to act. You may want to throw parts of this out entirely. You may find yourself drawn to certain ideas and pushing others away.

All of those impulses make sense. And if you are feeling some or all of this, you are exactly in the right place.

The world you are operating in is optimized for translation into action. It will tell you that this is the moment to define the problem, choose a metric, assemble a plan, and start building against familiar reference points. Insight, after all, is valuable only if it becomes execution.

Before you do that, I want to ask something different of you. Climb out of the stew, just for a moment, and observe. Think of this not as withdrawal but as orientation. This pause allows you to reflect on what a third-space moonshot actually requires in a world organized around speed, legibility, and validation through existing metrics.

A third-space moonshot is not built by following the loudest signals. It asks you to go where most people don't, can't, or won't—not out of contrarianism but because that is where undervalued needs, overlooked intelligence, and unrealized opportunity tend to gather.

The scale of the challenges we face and the possibilities before us now exceed our inherited imagination. Meeting them requires more. It requires expanding our capacity to perceive, to listen, and to hold ambiguity long enough for something genuinely new to reveal itself and then collectively discover, translate, build, integrate, evolve, cascade, benefit, and then go again.

So let's pause—not to decide but to breathe in what we've been journeying through. Pause to grapple.

The Discipline of Grapple

Grappling is not confusing. And it is not paralysis. The grappling required here is not a posture of hesitation or slowness. It is a commitment to a different temporal scale—to complexity, not because complexity is fashionable but because it carries its own clarity when we stay with it as long as required. This is not about making the work heavier or more difficult for its own sake. It is about moving with intention toward difference, not just toward whatever technological solution happens to be buildable.

Grappling is the disciplined refusal to collapse complexity too early.

In environments shaped by urgency, clarity is often mistaken for readiness. But clarity that arrives too quickly is usually borrowed from precedent, from metrics, from categories that already exist. Grappling interrupts that reflex. It creates space for orientation to form before execution hardens into path dependence.

Moonshots are not point A to point B experiences. They are inherently messy. They move through uncertainty, contradiction, and revision. Being in the back-and-forth is not a failure of leadership or vision; it is often the condition that allows real breakthroughs to take shape.

What Grapple Surfaces

When we allow ourselves to stay in the grapple, certain things begin to surface—often uncomfortably.

What we are optimizing for: grappling with complex problems has a way of revealing what we have been optimizing for, often without realizing it.

Speed feels neutral until you notice what it leaves behind.

Metrics feel objective until you ask what they fail to capture.

Scale feels virtuous until you see what abstraction erases.

These optimizations are not wrong. But they are not neutral either. They shape what becomes visible, what gets resourced, and what quietly falls away.

Who we think we are as builders: for many people, grappling destabilizes identity. The part of us that knows how to move quickly, make decisions, and deliver outcomes suddenly feels insufficient. Expertise becomes less portable.

Confidence becomes quieter. This can feel disorienting especially for high-capacity leaders who are used to decisiveness. But this discomfort is not a sign that something has gone wrong. It is often a sign you are working at a scale where old identities no longer fully apply.

I was once working with a company valued at more than $100 billion. They had begun to feel the existential threat of AI. Everything felt like it was "now or never." The leadership felt that if they didn't adopt and adapt immediately, they would lose everything. One of the VPs wanted to forge forward with a new path to expanding opportunities to create a new pipeline of moonshots.

They didn't want to grapple; they wanted to "find and go." But even though every trained instinct told them to choose and move, they instead chose to grapple. I asked them to spend six months in a state of radical exploration—not just with internal staff but with the distributed workers who had real-time insight into new pathways.

Those conversations were tough because they came up against a massive bias to action. But I told them: if you want new business lines, you can find them. If you want a moonshot, you have to discover them. That discovery comes only from the very real choice to stay in the grapple when the world is screaming at you to run.

Where power actually sits: grappling also exposes where power has been hiding—not just in institutions, but in timelines, incentives, and default expectations about what "progress" looks like.

- Who gets to wait?
- Who is forced to move quickly?
- Who defines value?

- Who absorbs risk when things don't work?
- Who defines the problem that needs to be solved?

These questions rarely surface when we rush. They become visible when we slow down just enough to see the system speaking back.

The "Nervous System" Thought Experiment

Now, not as a declaration, but as an exercise, imagine you choose to pursue a third-space moonshot. Not yet what it is or how it works or how it scales—but the orientation it would require.

Imagine you are looking at a problem that everyone else sees as a single, urgent event. A breach to be stopped. A gap to be filled. A threat to be neutralized. The world is screaming for a shield.

Now, Imagine You Refuse to Build the Shield

Instead, you stay in the grapple long enough to see that the "breach" is just a symptom of a silent landscape. You realize that what is actually missing isn't a better defense but a nervous system for the environment where that breach happened.

What if your moonshot wasn't the tool that reacts to the crisis but the intelligence layer that makes the entire ecosystem legible for the first time? What would you have to listen to differently if you were no longer looking for the "enemy" but mapping the "flow"?

You would have to resist the loud signal of the immediate threat to hear the quiet, constant pulse of the system itself. You would have to bring different forms of intelligence into relationship—weather, movement, history, silence—that the market might treat as irrelevant.

In this posture, you aren't just solving a problem. You are unlocking a new dimension of possibility. You are moving from building a product that "stops" something to building an infrastructure that "starts" something else entirely.

Don't picture a sleek interface or a rocket launch. Picture a rural supply chain that suddenly breathes because an "undervalued" bit of local data bridged the gap between a small-scale farmer and a global market. Picture a neighborhood where the "waste" of one industry becomes the "energy" for the next, not because of a new machine but because you built the connective tissue that allowed them to see each other. Or imagine:

- A global skill-wallet that replaces degrees with real-time, blockchain-verified competency logs earned through building real-world solutions
- Digital material tag that identifies the chemical value of steel and timber inside standing buildings, making demolition a commodity harvest
- A data provenance ledger for folk-art that uses acoustic and visual AI-fingerprinting to protect and monetize local artistic lineages in the metaverse

To get here, what signals would you have to resist? What kinds of intelligence would need to be brought into a relationship? What assumptions about speed, ownership, and control would need to loosen?

This is an ecosystem shift. It is the choice to let opportunity reveal itself through collective intelligence rather than imposing it through design. Technology becomes the co-partner—not a solution in search of a problem but the bridge to a new possibility. This does not describe a guaranteed outcome. It describes a posture. A way of becoming scaled possibility.

Choice

Choice is the pulse knocking at our door. Not choice as preference. Not choice as optimism. But the choice to be present enough to listen long enough for an ecosystem to tell us what it actually needs and how it needs it.

The choice to shape moonshots that don't merely insert technology into a system but alter something fundamental within it. Moonshots that, through collective intelligence and ingenuity, create cascading ripples of impact—ripples that unlock new paths of income, health, stability, and flourishing. Not for one. Not for thousands. But for millions. Even hundreds of millions of realities.

This is not a motivational speech meant to convince you that you can do anything. It is something more demanding. You have to want this. You have to believe it is possible. And then you have to be willing to hold what that belief asks of you.

When We Don't Choose

I once found myself standing in a parking lot outside one of Alphabet's Other Bets companies, talking with a moonshot leader who had poured years of his life into the work.

"So, Salima," he asked, "how do you think we're doing building this moonshot?"

I let the silence sit longer than was comfortable.

"It's 2023," I finally said, "and we are still skeptically asking whether 90% of the world's needs, geographies, and people are of value. How do you think we're doing?"

My question wasn't confrontational. It was genuinely ironic. Somewhere along the way, we shrank our sense of what is possible and who is possible. We became careful where we needed to be bold. Efficient where we needed to be expansive. Safe where we needed to be audacious.

Nonchoice does not keep things neutral. It selects by default. When we don't choose to ignite ecosystem-level opportunities, something else always does. When dominant trends or prevailing modes of invention become our reference point, we collapse into building in response rather than building in unimagined possibility. The world stops appearing as an open field of opportunity and begins to register as a narrow signal—loud in a few places, silent everywhere else.

Markets, as they currently operate, start to speak on behalf of value. They tell us what matters. What is investable. What is worth attention. Over time, we begin to see value only where it is already validated, already capitalized, already named. This is how possibility quietly contracts.

The Next Act

The world will continue to change. That is inevitable. The question is not whether this tension will exist. It already does and will continue to do so. The question is whether we will meet it with the 10X'd 10X audacity the world needs.

Third-space moonshots do not emerge accidentally. They require people willing to sit inside friction without outsourcing responsibility upward to institutions or downward to individuals. People willing to hold complexity without collapsing into cynicism or fantasy.

If this feels harder than expected, that's not a sign you're doing it wrong. It's a signal that you're finally working at the scale the moment demands.

And from here, the path forward is no longer abstract. It is a choice.

Moonshot Dojo

We've arrived at a point in the practice where to enable a third-space moonshot, we have to ask ourselves:

What moves will I make—and what moves will I outcreate?

What moves will I not choose?

How will I allow the world's ecosystems to move me—move us?

What must we choose—so that something unprecedented can emerge?

(Generated with AI using Google Gemini)

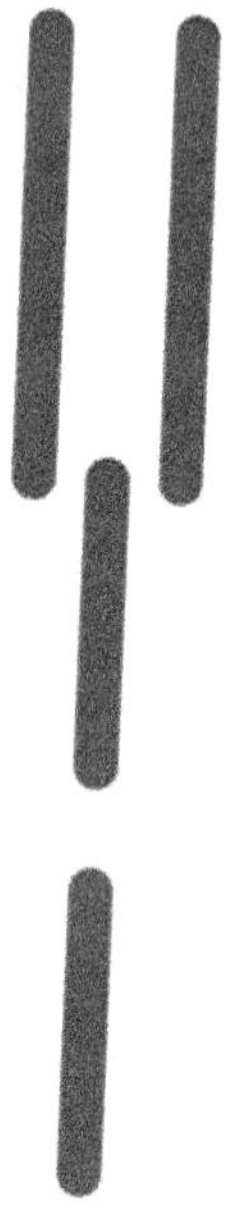

CHAPTER

9

Something More, Something Other

Take a beat.

Step outside, if you can. Somewhere you can see the horizon. Before you read on, listen to "Lucy in the Sky with Diamonds" by the Beatles.

Let the unexpectedness of the words do its work. Familiar objects slip out of their usual frames. Perception loosens—not to escape the world but to see it differently. This shift is not accidental. It is part of what this book has been moving toward.

The world is what it is. But it is also more than it is, simultaneously.

Nothing has changed. And yet, everything is slightly re-patterned.

That is the terrain you are standing in now.

I've asked you to listen to music throughout this journey because the third space cannot be accessed through optimization alone. "Lucy in the Sky with Diamonds" is not background sound. It is an exercise in relational perception. It forces the mind to hold two images at once—the mundane and the miraculous.

To build what comes next, you must be able to see the world as it is and as it could be, at the same time.

The Weight and the Fire

There is a weight that comes with building at this scale.

I felt it every day working with the Other Bets companies at Alphabet. I would walk into the colorful halls of Google, pass the micro-kitchens, and pause in the foyer. I sat in the stillness long enough to absorb the gravity of the place.

It was fun.

It had prestige and clout.

People assumed importance simply because of the badge.

But the truth is, I didn't work there. I built there. I pressed. I pushed. I expanded. I presented the opportunity for people to leap—every day.

If the heart of a moonshot is to build something of epic proportions—technologically and monetarily, for millions—then it must fundamentally change how life is lived. And if it has the power to do that, the world does not remain abstract or silent. It sits on your shoulders, tapping, insisting that *we* do what could not otherwise be done.

One day, I was sitting by the windows of the Rails building when a leader from an early pipeline project sat beside me. We stayed quiet for a long time. Eventually, she said, "Gosh, it's mighty lonely up here."

This wasn't the loneliness people usually mean when they talk about leadership. It was something else. To live in the audacity of changing the lives of millions is not a common path. It is lonely because few are willing to move in that direction.

The Act of Birth

At X, there was a ritual led by Maya Rudan where we mourned moonshot projects that had died—timely and untimely deaths. Alongside those, we mourned the people we had lost.

As I listened to eulogies for failed code and closed projects, one truth kept returning to me: letting go is not just an ending. It is an act of birth.

I was 5 years old when my older brother—12 at the time—asked me to help him with a recording session. He didn't write anything down. He spoke into the microphone about galaxies, planets, and universes talking to one another. He talked about traveling to faraway places to grow and learn.

In the background, he cued me to make space sounds—the ee-oohs and swooshes—in exactly the right moments.

Halfway through, he looked at me and said, "But to do this, I will have to leave you. Don't be sad when I am gone. I will miss you deeply. Just know that I have gone on the journey I am meant for."

A few months later, he died in a car accident.

Decades later, I was sitting with Tekedra Mawakana, the co-CEO of Waymo. We were talking about mission, math, and the grueling work of building self-driving technology. She looked at me and said, "Salima, do you realize that your personal story of loss is connected to Waymo's purpose?"

In that moment, the Rails and the recording session became one. A billion-dollar moonshot collided with the grounded loss of a 5-year-old girl in Toronto, Canada.

It wasn't about robots. It was about the sanctity of the journey.

That loss became an undeniable fire in me. I cannot live accepting only what is. The world is pulsing with what could be. Being present enough to feel that pulse is what makes the next leap in moonshots possible.

An Ecology

The world is not a puzzle to be solved. It is an ecology to be inhabited.

When we build from the top down, we sometimes treat life like a mathematical problem. When we build from the ground up, we sometimes treat it like a survival map. But when we build in the third space, we acknowledge something far more radical: we are stepping into unknown potential.

I felt the weight at Alphabet because I knew every line of code we wrote could be a new rusted switch in someone's life and that "millions of lives" are not simply data points.

They are people boiling milk in cupboards.

They are children making space sounds into a tape recorder.

If you feel the weight, that is appropriate.

That weight is accountability for something more—something other.

From the 16th Beat to the Opening

Recently, I sat with my mother as the afternoon light shifted across the room. She spoke casually about the world she once lived in—a world that feels distant now because it is.

She described washrooms that were once holes in the ground. Fires built from sticks and coal. Milk boiled in the morning and kept in cupboards to last the day.

Now she lives in a world of talking ovens and cartons from the grocery store. She held her iPad as she spoke, laughing at the absurdity of an 82-year-old woman scrolling through social media.

"So much has changed," she said. "In many good ways."

Then she paused. Lowered the iPad. Looked past the screen. And said a version of this in Gujarati:

> *What has always mattered is that people experience themselves. More of life.*
>
> *When these changes are right, they don't distract us from who we are. They make us more human.*

The room went quiet.

Because the work ahead is not about rushing forward or following a single story of progress. It is about staying present with what is—our lives, our systems, our world—while remaining open to what else is possible.

The technology is the lever.

Humanity is the moon.

The world is unfinished.

As are we.

Afterword

For Those Who Choose This Path

From Fire to Mandate

In the first few chapters of this book, I said that this is not about choosing a lineage or path of invention. It is about seeing that you have a repertoire in front of you.

Alphabet set the road in one way. We've dived into a Third Space. Our purpose is to bring forth a new orientation—a new set of possibilities in the face of a world that is screaming at us for other, more, different.

Yes, you have to be willing to walk on the wild side. You must move outside of predetermined or inherited logics about how invention happens, where value lives, and who can build the path to new ways of evolving human experience.

Remember, the third-space moonshot is a challenge to:

- **Discover and unlock** unseen and undervalued needs and opportunities across and within ecosystems
- **Use technology as a co-partner,** placing it at the highest value points to enable what would otherwise be impossible without it

- **Trigger cascading outcomes** across systems and relationships that benefit millions
- **Activate undervalued markets** by changing the very conditions of value creation
- **Unlock durable flows** of economic and societal benefit and value at the most critical points in an ecosystem

With this mandate, we are not just playing in the possible; we are playing in what might initially feel impossible. That is the point. It should feel like you can't fully see it yet, but you are absolutely committed to discovering and building it—not for the world but with the world.

It's slightly crazy. And if you feel that, good. Now we are getting somewhere.

This Is Not an App

I want to be blunt as we close this.

The world has people who want to build incremental optimization. It has people who will build platforms that shave three seconds off a delivery or a half-cent off a transaction. And we will continue to need those kinds of inventions, products, and solutions.

This—this third-space moonshot we've been grappling with—is not a comfortable road.

If your "moonshot" doesn't make you tremble a little at the audacity of its reach, then you are still coloring within the lines. If it doesn't require you to hold the dynamic reality of the street and the galactic potential of technology simultaneously, then you are not building a third-space moonshot.

Go build an app if you want a business. I've got plenty of those ideas myself.

But if you want unprecedented impact—if you want to be an architect of human experience and expansion—then you must go where the logic breaks. Go where the "standard" fails. Go to the Masindi Ports of the future, to the rinks-turned-malls, to the unpaved tracks where the current of unlimited potential is actually flowing.

The Repertoire: The Mechanics of the Third Space

I hope you walk away from this book wanting to wrestle for a moonshot, rather than settling for building one.

This is not a point A to B road. It is an orchestration.

Before you apply the mandate, you must adopt these mechanics:

- **Believe**

  You can and should create ecosystem-level solutions that hundreds of millions will reap the benefits from. If your vision doesn't aim for this scale, it's a product, not a moonshot.

- **Trust the Signal (The Beaver)**

  Stop over-relying on linear roadmaps. Follow the signals from the ecosystem on what must be built, not just what is buildable.

- **Orchestrate, Don't Solo (Paganini)**

  Refuse the "solo genius" myth. Deliberately build with the multiple intelligences and worlds of Nairobi, Chennai, Tokyo, São Paulo, Dubai, Berlin, and Chicago;

with the experience of the engineer, the elder, the gig worker, the nurse, and the scientist—until you collectively unlock what was previously unreachable.

- **Discern the Tool (The Hygienist)**

  Shape and place technology as a co-partner. If the tool doesn't reach a place the human hand cannot, it's not the moonshot partner.

- **Flip the Rusted Switch**

  Stop looking for a "buyer." Look for the misaligned relationship, the undervalued places and opportunities, or the trust deficit. Enable new flows, and the market will emerge.

- **Use the Momentum (Intellectual Aikido)**

  Don't fight the inevitable frictions that others run from; use that energy to shape the trajectory of the moonshot.

- **Out-Design**

  Refuse the "static plug and play." Constantly ask: What else is possible? Who does this benefit? Who does it harm? Design the guardrails into the architecture itself.

- **Stay in the Grapple**

  Refuse to collapse complexity for the sake of a quick metric. Stay in the mess until the nervous system of the solution reveals itself.

- **Risk**

  If the current metrics and traditional signals are keeping you from leaping, it's not a third-space moonshot. If you don't feel a fire in your belly, it isn't audacious enough.

The Choice

You can create anything. You create every day. At varying degrees of scale and impact.

We often first see what we don't have; it's a well-trained instinct. In the world of invention, we start by looking at what already exists. But we don't sit in either place. We live in the space where possibility and constraint copulate. And in that friction, there is a force.

Remember my hypothesis from the start of this journey: that we have been too slow, too insulated, and too safe.

Now, we flip it.

We move at the temporal speed the world requires. We incubate with the world. And we stand in the greatest risk of our lives—that we built something of unprecedented impact and scale.

So inevitably, the force is not to create a product or a solution. The force is to architect something that creates new pathways. That reorients what's already there. That shifts trajectories and outcomes—yours, mine, ours—economically, socially, culturally, technologically.

For too long, we have been letting others decide because we've seen how power becomes concentrated. But the greatest power is collective ingenuity, not solo rocket tests.

Propulsion into the future is not a one-person game. It is an orchestration.

If you feel resistant to this vision, this mandate, that's good. Ask yourself why. What does it push up against? What is it quietly—or loudly—nudging open?

Some of you may work through all of that and decide, "*Nah. I'll stick to this or that playbook.*" And others of you may find yourself more awake and ready for this.

At the end of the day, this is not a business, invention, or tech playbook.

It is something meant to embed itself in your psyche, in your perception, in your actions, and in your choices.

Maybe you'll take this to your teams, your community, or your company.

Maybe you will say, "I am ready to launch a full-scale third-space moonshot."

What matters is that you now feel another way of shaping the next 5, 10, 15, 20 years. This orientation will stay with you, long after you've left this page.

"Faultlines": Nusrat Fateh Ali Khan, Michael Brooks

Acknowledgments

This book reflects the many people, contexts, and experiences across Alphabet, technology, regions, communities, and unnamed places that have shaped how I choose the creative impulse—what feels needed and what feels necessary.

Thank you to my editors at Wiley, specifically, Jim Minatel, Annie Melnick, Caroline Define, Kim Wimpsett, and Moses Ashirvad. Your encouragement is fuel.

To my husband, Colin, mom, Jubeda, my dad, Noorali—thank you for the thousands of yeses.

And to my son, Oscar, who at three years old told me, "Mommy, don't prepare. Propel."

You are the true spirit of possibility.

Propel.

Notes

Introduction

1. Angel One. (2025). *Gujarat Startups Raise Nearly ₹400 Crore Since April as Funding Revival Gains Momentum*. https://www.angelone.in/news/startups/gujarat-startups-raise-nearly-400-crore-since-april-as-funding-revival-gains-momentum
2. Nogel, R. J. (1973). *Human rights and Uganda's expulsion of its Asian minority*. Denver Journal of International Law and Policy, 3,219–247.https://digitalcommons.du.edu/cgi/viewcontent.cgi?article=2135&context=djilp
3. Songwriters Hall of Fame. (n.d.). *Fly Me to the Moon*. https://www.songhall.org/profiles/fly-me-to-the-moon
4. Wikipedia contributors. (n.d.). *Planetarium*. https://en.wikipedia.org/wiki/Planetarium
5. UNESCO World Heritage Centre. (n.d.). *Jantar Mantar, Jaipur*. https://whc.unesco.org/en/list/1338
6. Teller, A. (2024). *A conversation with Alphabet's captain of moonshots*. The Wall Street Journal Podcast. https://www.wsj.com/podcasts/the-journal/a-conversation-with-alphabets-captain-of-moonshots/b357d2d9-edd6-47e4-9a72-5c26a3a62e9a
7. NBC News. (2025). *Watch: Waymo Driverless Robotaxi Sends Airport Passenger Spinning in Circles*. https://www.nbcnews.com/news/us-news/watch-waymo-driverless-robotaxi-sends-airport-passenger-spinning-circl-rcna186672
8. CompaniesMarketCap. (n.d.). *Largest Companies by Market Capitalization*. https://companiesmarketcap.com
9. Page, L. & Brin, S. (2015). *Alphabet Founder's Letter*. https://abc.xyz/investor/founders-letters/2015/default.aspx

Chapter 1

1. de Laet, M. & Mol, A. (2000). The Zimbabwe Bush Pump: Mechanics of a fluid technology. *Social Studies of Science, 30*(2), 225–263.https://www.trans-techresearch.net/wp-content/uploads/2021/10/de-Laet-and-Mol-The-Zimbabwe-Bush-Pump-.pdf
2. Rural Water Supply Network. (2013). *How Three Handpumps Revolutionised Rural Water Supplies: The Zimbabwe Bush Pump.* https://rwsn.blog/2013/03/21/how-three-handpumps-revolutionised-rural-water-supplies-the-zimbabwe-bush-pump/
3. X (Moonshot Factory). (2017). *From Mall to Moonshot Factory: The History of X's Home Turf.* https://blog.x.company/from-mall-to-moonshot-factory-the-history-of-xs-home-turf-80d61b32b73a
4. Search Engine Land. (2025). *Google Now Sees More Than 5 Trillion Searches per Year.* https://searchengineland.com/google-5-trillion-searches-per-year-452928
5. AlphaFold Protein Structure Database. (n.d.). *AlphaFold DB.* https://alphafold.ebi.ac.uk
6. GV. (n.d.). *About GV.* https://www.gv.com/about
7. CapitalG. (n.d.). *About CapitalG.* https://capitalg.com/about/
8. Waymo. (2025). *Delivering More for Our Riders in a Year of Incredible Growth (2025 Year in Review).* https://waymo.com/blog/2025/12/2025-year-in-review
9. Waymo. (2025). *Waymo 2025 Tourism Impact Report* [PDF]. https://storage.googleapis.com/waymo-uploads/files/site-media/Blog/Waymo%202025%20Tourism%20Impact%20Report.pdf
10. Google DeepMind. (2016). *DeepMind AI Reduces Google Data Centre Cooling Bill by 40%.* https://deepmind.google/blog/deepmind-ai-reduces-google-data-centre-cooling-bill-by-40/
11. Google DeepMind. (2019). *Machine Learning Can Boost the Value of WindEnergy.*https://deepmind.google/blog/machine-learning-can-boost-the-value-of-wind-energy/
12. Intrinsic. (n.d.). *About Intrinsic.* Alphabet Inc. https://intrinsic.ai
13. International Federation of Robotics. (2024). *World Robotics Report* (market size and growth projections).https://ifr.org/worldrobotics

Chapter 2

1. Ingle, S. (2018). *'Creative' AlphaZero Leads Way for Chess Computers and, Maybe, Science.* The Guardian. https://www.theguardian.com/sport/2018/dec/11/creative-alphazero-leads-way-for-chess-computers-and-maybe-science

Chapter 3

1. Gojek Official Site Gojek. (n.d.). *Gojek*. https://www.gojek.com/en-id
2. Gojek Economic Impact Report Page Gojek. (2024). *Ekonomi Digital: Digital Economy Contribution to Indonesia* (impact summary). https://www.gojek.com/blog/gojek/ekonomi-digital?utm_source
3. Estonia Digital Government Overview e-Estonia. (2024). *Estonia Is at the Top of the UN e-Government Ranking*. https://e-estonia.com/estonia-is-at-the-top-of-the-un-e-government-ranking/
4. Artha Global report (PDF) Artha Global. (2025). *The Economic Impact of Digital Public Infrastructure: From Hype to Reality (Paper 3)*. https://artha.global/wp-content/uploads/2025/11/The-Economic-Impact-of-DPI_-From-Hype-to-Reality-Paper-3.docx.pdf
5. TEDx Talks. (2014). *Space: The Ultimate Startup | Chris Boshuizen | TEDxSanFrancisco* [Video]. YouTube. https://www.youtube.com/watch?v=wp2sHc_PKjc
6. Planet Labs Official Site Planet Labs PBC. (n.d.). *Planet*. https://www.planet.com/
7. Marshall TED Talk (2014). Marshall, W. (2014). *Tiny Satellites show us the Earth as it Changes in near-real-time* [Video]. TED Conferences. https://www.ted.com/talks/will_marshall_tiny_satellites_show_us_the_earth_as_it_changes_in_near_real_time
8. Schingler keynote (2018). Schingler, R. (2018). *Eyes in the Sky, Feet on the Ground* [Speech]. Founders Pledge Forum, London, United Kingdom. YouTube. https://www.youtube.com/watch?v=aDs_qYiPWFI

Chapter 4

1. Banerjee, S. (2010). *Harish Hande of SELCO India: Shedding Light on India's Underserved Markets*. Knowledge@Wharton, University of Pennsylvania. https://knowledge.wharton.upenn.edu/article/harish-hande-of-selco-india-shedding-light-on-indias-underserved-markets/
2. Knowledge at Wharton Staff. *Harish Hande of SELCO India*.
3. Yale School of Management Case Research and Development Team. (n.d.). *SELCO* (Case No. 7648). Yale School of Management. https://cases.som.yale.edu/node/7648
4. Lothe, R. (2012). *Harish Hande Providing Solar Power to the People in India*. Bellona Foundation. https://bellona.org/news/renewable-energy/solar/2012-02-harish-hande-providing-solar-power-to-the-people-in-india

5. Reykjavík Energy. (n.d.). *The History of Geothermal Heating in Reykjavík*. https://www.orkuveita.is/en/about-us/history/
6. Landsvirkjun. (n.d.). *Bjarnarflag Power Station: Iceland's First Geothermal Plant*. https://www.landsvirkjun.com/powerstations/bjarnarflag
7. Logadóttir, H. H. (2013). Iceland's sustainable energy story: A model for the world? *UN Chronicle*, *52*(3), 40–45. https://www.un.org/en/chronicle/article/icelands-sustainable-energy-story-model-world
8. Yang, J. (2022). An empirical study evaluating the symbiotic efficiency of China's provinces and the innovation ecosystem in the high-tech industry. *Complexity*, 2022. https://doi.org/10.1155/2022/1391415
9. Leong, T. Y. (1998). Multiple perspective dynamic decision making. *IEEE Intelligent Systems*, *13*(6), 44–55. https://ieeexplore.ieee.org
10. Rahayu, H. P., Comfort, L. K., Haigh, R., Amaratunga, D., & Khoirunnisa, D. (2020). A study of people-centered early warning system in the face of near-field tsunami risk for Indonesian coastal cities. *International Journal of Disaster Resilience in the Built Environment*, *11*(4), 497–512. https://doi.org/10.1108/IJDRBE-10-2019-0068
11. Knowlton, K., Kulkarni, S. P., Azhar, G. S., Mavalankar, D., Jaiswal, A., Connolly, M., Nori-Sarma, A., Rajiva, A., Dutta, P., Deol, B., Sanchez, L., Khosla, R., Webster, P. J., Toma, V. E., Sheffield, P., & Hess, J. J. (2014). Development and implementation of South Asia's first heat-health action plan in Ahmedabad (Gujarat, India). *International Journal of Environmental Research and Public Health*, *11*(4), 3473–3492. https://doi.org/10.3390/ijerph110403473
12. Open Insulin Foundation. (n.d.). *Who We Are*. https://openinsulin.org/who-we-are/
13. Gallegos, J. E., Boyer, C., Pauwels, E., Kaplan, W. A., & Peccoud, J. (2018). The open insulin project: A case study for 'biohacked' medicines. *Trends in Biotechnology*, *36*(12), 1211–1218. https://doi.org/10.1016/j.tibtech.2018.07.009

Chapter 5

1. Kemp, S. (2025). *Digital 2026 global overview report*. We Are Social; Meltwater. https://wearesocial.com/me/blog/2025/10/digital-2026-global-overview-report/
2. Alation. (2025). *What Is Dark Data? Complete Guide and Best Practices*. https://www.alation.com/blog/dark-data/

3. Wahlroos, M., Pärssinen, M., Rinne, S., Sykes, S., & Manner, J. (2018). Future views on waste heat utilization – Case of data centers in Northern Europe. *Renewable and Sustainable Energy Reviews*, *82*, 1749–1764. https://doi.org/10.1016/j.rser.2017.10.058
4. Splunk. (2024). *The State of Dark Data: The Latent Value of Machine-generated Intelligence*. Splunk Inc. https://www.splunk.com/en_us/form/the-state-of-dark-data.html

Chapter 6

1. McKinsey & Company. (2025). *The New Face of Wealth: The Rise of the Female Investor*. McKinsey & Company. https://www.mckinsey.com/industries/financial-services/our-insights/the-new-face-of-wealth-the-rise-of-the-female-investor
2. Oppenheimer & Co. Inc. (2025). *Redefining Women and Wealth: A Financial Revolution*. Oppenheimer. https://www.oppenheimer.com/news-media/2025/insights/articles/june/redefining-women-and-wealth-a-financial-revolution?utm_source

Chapter 7

1. Khaldun, I. (1967). *The Muqaddimah: An Introduction to History* (F. Rosenthal, Trans.). Princeton University Press. (Original work published 1377).